T0192800

Newtonian Dynamics

Newtonian Dynamics
An Introduction

Richard Fitzpatrick

CRC Press
Taylor & Francis Group
Boca Raton London New York

CRC Press is an imprint of the
Taylor & Francis Group, an **informa** business

First edition published 2022
by CRC Press
6000 Broken Sound Parkway NW, Suite 300, Boca Raton, FL 33487-2742

and by CRC Press
2 Park Square, Milton Park, Abingdon, Oxon, OX14 4RN

CRC Press is an imprint of Taylor & Francis Group, LLC

Library of Congress Cataloging-in-Publication Data

Names: Fitzpatrick, Richard, 1963- author.
Title: Newtonian dynamics : an introduction / Richard Fitzpatrick.
Description: First edition. | Boca Raton : CRC Press, 2022. | Includes
  bibliographical references and index.
Identifiers: LCCN 2021031651 | ISBN 9781032046624 (hardback) | ISBN
  9781032056661 (paperback) | ISBN 9781003198642 (ebook)
Subjects: LCSH: Dynamics--Textbooks.
Classification: LCC QA845 .F58 2022 | DDC 531/.11--dc23
LC record available at https://lccn.loc.gov/2021031651

ISBN: 978-1-032-04662-4 (hbk)
ISBN: 978-1-032-05666-1 (pbk)
ISBN: 978-1-003-19864-2 (ebk)

DOI: 10.1201/9781003198642

Publisher's note: This book has been prepared from camera-ready copy provided by the authors.

Contents

Preface

The aim of this book is to give a concise, comprehensive, college-level introduction to the branch of physics known as Newtonian dynamics. The book assumes a high-school-level knowledge of physics, as well as a familiarity with geometry, trigonometry, algebra, and calculus.

Newtonian dynamics is the study of the motion of bodies (including the special case in which bodies remain at rest) in accordance with the general principles first enunciated by Sir Isaac Newton in his *Philosophiae Naturalis Principia Mathematica* (1687), commonly known as the *Principia*. Newtonian dynamics was the first branch of physics to be discovered, and is the foundation upon which all other branches of physics are built. Moreover, Newtonian dynamics has many important applications in other areas of science, such as astronomy (e.g., celestial mechanics), chemistry (e.g., the dynamics of molecular collisions), geology (e.g., the propagation of seismic waves, generated by earthquakes, through the Earth's crust), and engineering (e.g., the equilibrium and stability of structures). Newtonian dynamics is also of great significance outside the realm of science. After all, the sequence of events leading to the discovery of Newtonian dynamics—starting with the ground-breaking work of Copernicus, continuing with the research by Galileo, Kepler, and Descartes, and culminating in the monumental achievements of Newton—involved the complete overthrow of the Aristotelian picture of the universe, which had previously prevailed for more than a millennium, and its replacement by a recognizably modern picture in which the Earth and humankind no longer play a privileged role.

The central message of the *Principia* is that the same laws of motion that govern the movements of everyday objects on the surface of the Earth also govern the movements of celestial bodies. Hence, the first eleven chapters of this book are devoted to the application of Newton's laws of motion to the dynamics of everyday objects, whereas the final four chapters employ the same laws to analyze the dynamics of celestial bodies. In particular, the final section of the book covers the following important topics: the orbits of artificial satellites around the Earth; the orbits of the planets around the Sun; the rotational flattening of the Earth; lunar and solar tides; and the precession of the equinoxes.

This book starts off at a mathematical level that is suitable for beginning undergraduates. The level of the mathematics gradually increases, as the book proceeds, until it (almost) reaches that of a typical upper-division physics course. Hence, this book constitutes an ideal vehicle for easing the difficult transition between elementary and advanced undergraduate physics courses.

A large number of exercises are included at the end of each chapter. Some of these exercises involve simple numerical calculations, while others require more extensive mathematical analysis. All of the exercises are accompanied by their final answers (when appropriate). The aim of the exercises is to allow the student to accurately gauge their understanding of the material presented in the book.

Acknowledgments

The material presented in Section 3.2 of this book is reproduced from Sections 2.2, 2.4, and 2.5 of Fitzpatrick 2008. (courtesy of Jones & Bartlett).

The material presented in Section 5.7 of this book is reproduced from Section 1.7 of Fitzpatrick 2012. The material presented in Sections 15.2, 15.3, 15.4, and 15.5 of this book is reproduced from Chapter 2 of Fitzpatrick 2012. Most of the material presented in Chapter 14 of this book is reproduced from Chapter 3 of Fitzpatrick 2012. The material presented in Sections 12.2, 12.3, 12.4, 15.6, and 15.7 of this book is reproduced from Chapter 5 of Fitzpatrick 2012. Finally, most of the material presented in Section 15.8 of this book is reproduced from Section 7.10 of Fitzpatrick 2012 (courtesy of Cambridge University Press).

The material presented in Sections 11.2, 11.4, and 11.5 of this book is reproduced from Chapter 1 of Fitzpatrick 2019 (courtesy of Taylor & Francis).

Measurement and Units

1.1 MKS UNITS

The first principle of any exact physical science is measurement. In Newtonian dynamics, there are three fundamental quantities that are subject to measurement:

1. Intervals in space; that is, **length**.

2. Quantities of inertia, or **mass**, possessed by various bodies.

3. Intervals in **time**.

Any other type of measurement in Newtonian dynamics can (effectively) be reduced to some combination of measurements of these three quantities.

Each of the three fundamental quantities—length, mass, and time—is measured with respect to some convenient standard. The system of units currently used by most scientists and engineers is called the *mks system*—after the first initials of the names of the units of length, mass, and time, respectively, in this system. That is, the *meter*, the *kilogram*, and the *second*.

The mks unit of length is the **meter** (symbol: m). The meter was formerly the distance between two scratches on a platinum-iridium alloy bar kept at the International Bureau of Weights and Measures in Sèvres, France, but is now defined as the distance travelled by light in vacuum in $1/299\,792\,458$ seconds (Wikipedia contributors 2021).

The mks unit of mass is the **kilogram** (symbol: kg). The kilogram was formally defined as the mass of a platinum-iridium alloy cylinder kept at the International Bureau of Weights and Measures in Sèvres, France, but is now defined in such a manner as to make Planck's constant take the value $6.626\,070\,15 \times 10^{-34}$ when expressed in mks units (Wikipedia contributors 2021).

The mks unit of time is the **second** (symbol: s). The second was formerly defined in terms of the Earth's rotation, but is now defined as the time required for $9\,192\,631\,770$ complete oscillations associated with the transition between the two hyperfine levels of the ground state of the isotope Cesium 133 (Wikipedia contributors 2021).

In addition to the three fundamental quantities, Newtonian dynamics also deals with derived quantities, such as velocity, acceleration, momentum, angular momentum, etcetera. Each of these derived quantities can be reduced to some particular combination of length, mass, and time. The mks units of these derived quantities are, therefore, the corresponding combinations of the mks units of length, mass, and time. For instance, a velocity can be reduced to a length divided by a time. Hence, the mks units of velocity are meters per second:

$$[v] = \frac{[L]}{[T]} = \mathrm{m\,s^{-1}}. \tag{1.1}$$

DOI: 10.1201/9781003198642-1

Here, v stands for a velocity, L for a length, and T for a time, whereas the operator $[\cdots]$ represents the units, or dimensions, of the quantity contained within the brackets. Momentum can be reduced to a mass times a velocity. Hence, the mks units of momentum are kilogram-meters per second:

$$[p] = [M][v] = \frac{[M][L]}{[T]} = \text{kg m s}^{-1}. \tag{1.2}$$

Here, p stands for a momentum, and M for a mass. In this manner, the mks units of all derived quantities appearing in Newtonian dynamics can easily be obtained.

Some combinations of meters, kilograms, and seconds occur so often in physics that they have been given special nicknames. Such combinations include the newton, which is the mks unit of force, and the joule, which is the mks unit of energy. These so-called derived units are listed in Table 1.1.

1.2 STANDARD PREFIXES

MKS units are specifically designed to conveniently describe those motions that occur in everyday life. Unfortunately, mks units tend to become rather unwieldy when dealing with motions on very small scales (e.g., the motions of molecules) or on very large scales (e.g., the motions of stars in the Milky Way galaxy). In order to help cope with this problem, a set of standard prefixes has been devised that allow the mks units of length, mass, and time to be modified so as to deal more easily with very small and very large quantities. These prefixes are specified in Table 1.2. Thus, a *kilometer* (km) represents 10^3 m, a *nanometer* (nm) represents 10^{-9} m, and a *femtosecond* (fs) represents 10^{-15} s. The standard prefixes can also be used to modify the units of derived quantities.

1.3 OTHER UNITS

The mks system is not the only system of units in existence. Unfortunately, the obsolete cgs (centimeter-gram-second) system, and the even more obsolete fps (foot-pound-second) system, are still in use today, although their continued employment is now strongly discouraged in science and engineering. Conversion between different systems of units is, in principle, perfectly straightforward, but, in practice, a frequent source of error. Witness, for example, the loss of the Mars Climate Orbiter in 1999 because the Lockheed Martin engineers who designed and built its rocket engine used fps units whereas the NASA mission controllers employed mks units. Table 1.3 specifies the various conversion factors between mks, cgs, and fps units. Note that a pound is a unit of force, rather than mass. Additional non-standard units of length include the inch (1 ft = 12 in), the yard (1 ya = 3 ft), and the mile (1 mi = 5 280 ft). Additional non-standard units of mass include the ton (in the US, 1 ton = 2 000 lb; in the UK, 1 ton = 2 240 lb), and the metric ton (1 tonne = 1 000 kg).

Table 1.1 Derived Units

Physical Quantity	Derived Unit	Abbreviation	MKS Equivalent
Force	newton	N	m kg s^{-2}
Energy	joule	J	$\text{m}^2 \text{ kg s}^{-2}$
Power	watt	W	$\text{m}^2 \text{ kg s}^{-3}$
Pressure	pascal	Pa	$\text{m}^{-1} \text{ kg s}^{-2}$

Table 1.2   Standard Prefixes

Factor	Prefix	Symbol	Factor	Prefix	Symbol
10^{18}	exa-	E	10^{-1}	deci-	d
10^{15}	peta-	P	10^{-2}	centi-	c
10^{12}	tera-	T	10^{-3}	milli-	m
10^{9}	giga-	G	10^{-6}	micro-	μ
10^{6}	mega-	M	10^{-9}	nano-	n
10^{3}	kilo-	k	10^{-12}	pico-	p
10^{2}	hecto-	h	10^{-15}	femto-	f
10^{1}	deka-	da	10^{-18}	atto-	a

Finally, additional non-standard units of time include the minute ($1\,\mathrm{min} = 60\,\mathrm{s}$), the hour ($1\,\mathrm{hr} = 60\,\mathrm{min}$), the (solar) day ($1\,\mathrm{da} = 24\,\mathrm{hr}$), and the (Julian) year ($1\,\mathrm{yr} = 365.25\,\mathrm{da}$).

1.4   DIMENSIONAL ANALYSIS

As we have already seen, length, mass, and time are three fundamentally different entities that are measured in terms of three completely independent units. It, therefore, makes no sense for a prospective law of physics to express an equality between (say) a length and a mass. In other words, the example law,

$$m = l, \tag{1.3}$$

where m is a mass and l is a length, cannot possibly be correct. One easy way of seeing that Equation (1.3) is invalid (as a law of physics) is to note that this equation is dependent on the adopted system of units. That is, if $m = l$ in mks units then $m \neq l$ in fps units, because the conversion factors which must be applied to the left- and right-hand sides of the equation differ. Physicists hold very strongly to the maxim that the laws of physics possess objective reality. In other words, the laws of physics are equivalent for all observers. One immediate consequence of this maxim is that a law of physics must take the same form in all possible systems of units that a prospective observer might choose to employ. The only way in which this can be the case is if all laws of physics are dimensionally consistent. In other words, the quantities on the left- and right-hand sides of the equality sign in any given law of physics must have the same dimensions (i.e., the same combinations of length, mass, and time). A dimensionally consistent equation naturally takes the same form in all

Table 1.3   Conversion Factors Between the mks, cgs, and fps systems of Units. Here, g, ft, and lb are the Abbreviations for gram, foot, and pound, respectively.

$1\,\mathrm{cm}$	=	$10^{-2}\,\mathrm{m}$
$1\,\mathrm{g}$	=	$10^{-3}\,\mathrm{kg}$
$1\,\mathrm{ft}$	=	$0.3048\,\mathrm{m}$
$1\,\mathrm{lb}$	=	$4.448\,\mathrm{kg\,m\,s^{-2}}$
$1\,\mathrm{slug}$	=	$14.59\,\mathrm{kg}$

possible systems of units, because the same conversion factors are applied to both sides of the equation when transforming from one system to another.

As an example, let us consider what is probably the most famous equation in physics: that is, Einstein's mass-energy relation,

$$E = m c^2. \tag{1.4}$$

Here, E is the energy of a body, m is its mass, and c is the velocity of light in vacuum. The dimensions of energy are $[M][L^2]/[T^2]$, and the dimensions of velocity are $[L]/[T]$. Hence, the dimensions of the left-hand side are $[M][L^2]/[T^2]$, whereas the dimensions of the right-hand side are $[M]([L]/[T])^2 = [M][L^2]/[T^2]$. It follows that Equation (1.4) is indeed dimensionally consistent. Thus, $E = m c^2$ holds good in mks units, in cgs units, in fps units, and in any other sensible set of units. Had Einstein proposed $E = m c$, or $E = m c^3$ then his error would have been immediately apparent to other physicists, because these prospective laws are not dimensionally consistent. In fact, $E = m c^2$ represents the only simple, dimensionally consistent way of combining an energy, a mass, and the velocity of light in a law of physics.

The last comment leads naturally to the subject of *dimensional analysis*. That is, the use of the idea of dimensional consistency to guess the forms of simple laws of physics. It should be noted that dimensional analysis is of fairly limited applicability, and is a poor substitute for analysis employing the actual laws of physics. Nevertheless, it is occasionally useful.

For instance, suppose that a special effects studio wants to film a scene in which the Leaning Tower of Pisa topples to the ground. In order to achieve this goal, the studio might make a scale model of the tower, which is (say) 1 m tall, and then film the model falling over. The only problem is that the resulting footage would look completely unrealistic because the model tower would fall over too quickly. The studio could easily fix this problem by slowing the film down. But, by what factor should the film be slowed down in order to make it look realistic?

Although, at this stage, we do not know how to apply the laws of physics to the problem of a tower falling over, we can, at least, make some educated guesses as to the factors upon which the time, t_f, required for this process to occur depends. In fact, it seems reasonable to suppose that t_f depends principally on the mass of the tower, m, the height of the tower, h, and the acceleration due to gravity, g. In other words,

$$t_f = C m^x h^y g^z, \tag{1.5}$$

where C is a dimensionless constant, and x, y, and z are unknown exponents. The exponents x, y, and z can be determined by the requirement that the previous equation be dimensionally consistent. Incidentally, the dimensions of an acceleration are $[L]/[T^2]$. Hence, equating the dimensions of both sides of Equation (1.5), we obtain

$$[T] = [M]^x [L]^y \left(\frac{[L]}{[T^2]} \right)^z. \tag{1.6}$$

We can now compare the exponents of $[L]$, $[M]$, and $[T]$ on either side of the previous expression. These exponents must all match in order for Equation (1.5) to be dimensionally consistent. Thus,

$$0 = y + z, \tag{1.7}$$

$$0 = x, \tag{1.8}$$

$$1 = -2 z. \tag{1.9}$$

It immediately follows that $x = 0$, $y = 1/2$, and $z = -1/2$. Hence,

$$t_f = C\sqrt{\frac{h}{g}}. \tag{1.10}$$

Now, the actual tower of Pisa is approximately $100\,\mathrm{m}$ tall. It follows that because $t_f \propto \sqrt{h}$ (g is the same for both the real and the model tower), the $1\,\mathrm{m}$ high model tower would fall over a factor of $\sqrt{100/1} = 10$ times faster than the real tower. Thus, the film must be slowed down by a factor of 10 in order to make it look realistic.

1.5   EXPERIMENTAL ERRORS

All experimental measurements are subject to errors. Indeed, an experimental measurement in physics is not considered complete unless it is accompanied by an estimate of the error in the measurement.

Consider a measurement made of some physical quantity, x. We can express the result of such a measurement in the form

$$x = \overline{x} + \delta x. \tag{1.11}$$

Here, $\overline{x}$ is the *mean value* of the measurement. In other words, $\overline{x}$ is the value obtained when a very large number of independent measurements of x are performed, and the resulting values are averaged. δx represents the *error* in an individual measurement of x. Let $\langle \cdots \rangle$ represent the average value of the quantity enclosed in the angle brackets. By definition, $\langle x \rangle = \overline{x}$. Hence, averaging Equation (1.11), we deduce that $\langle \delta x \rangle = 0$. Let

$$\langle (\delta x)^2 \rangle = \sigma_x^2. \tag{1.12}$$

Here, σ_x is an estimate of the magnitude of the error in an individual measurement of x. In other words, such a measurement is likely to yield a value lying between $\overline{x} - \sigma_x$ and $\overline{x} + \sigma_x$, and unlikely to yield a value lying outside this range.

Consider measurements made of two independent physical quantities, x and y. Suppose that

$$z = x + y. \tag{1.13}$$

We can write

$$x = \overline{x} + \delta x, \tag{1.14}$$
$$y = \overline{y} + \delta y, \tag{1.15}$$
$$z = \overline{z} + \delta z, \tag{1.16}$$

where $\langle \delta x \rangle = \langle \delta y \rangle = \langle \delta z \rangle = 0$. It follows that

$$\overline{z} + \delta z = \overline{x} + \delta x + \overline{y} + \delta y. \tag{1.17}$$

Averaging the previous equation, we obtain

$$\overline{z} = \overline{x} + \overline{y}. \tag{1.18}$$

In other words, not surprisingly, if z is the sum of x and y then the mean value of z is the sum of the mean values of x and y. Taking the difference between the previous two equations, we deduce that

$$\delta z = \delta x + \delta y. \tag{1.19}$$

Hence,

$$(\delta z)^2 = (\delta x)^2 + (\delta y)^2 + 2\,\delta x\,\delta y. \tag{1.20}$$

Averaging the previous equation, we get

$$\langle (\delta z)^2 \rangle = \langle (\delta x)^2 \rangle + \langle (\delta y)^2 \rangle + 2\,\langle \delta x\,\delta y \rangle. \tag{1.21}$$

However, $\langle (\delta x)^2 \rangle = \sigma_x^2$, $\langle (\delta y)^2 \rangle = \sigma_y^2$, and $\langle (\delta z)^2 \rangle = \sigma_z^2$, where σ_z is the magnitude of the error in the measured value of z. If x and y are independent physical quantities then we would expect the errors in the measurements of these quantities to be completely uncorrelated. In other words, $\langle \delta x\,\delta y \rangle = 0$. Thus, the previous equation yields

$$\sigma_z^2 = \sigma_x^2 + \sigma_y^2. \tag{1.22}$$

This expression specifies how the errors propagate in Equation (1.13).

A similar calculation to that just performed reveals that if

$$z = x - y \tag{1.23}$$

then

$$\overline{z} = \overline{x} - \overline{y}, \tag{1.24}$$

$$\sigma_z^2 = \sigma_x^2 + \sigma_y^2. \tag{1.25}$$

Suppose that N independent measurements are made of some physical quantity, x. Let x_i be the result of the ith measurement, and let

$$X = \sum_{i=1,N} x_i. \tag{1.26}$$

A straightforward generalization of Equation (1.22) reveals that

$$\sigma_X^2 = N\,\sigma_x^2. \tag{1.27}$$

Here, we are assuming that the errors in different measurements are uncorrelated; that is, $\langle \delta x_i\,\delta x_j \rangle = 0$ when $i \neq j$. We are also assuming that $\langle (\delta x_i)^2 \rangle = \sigma_x^2$ for all i. Let

$$\tilde{x} = \frac{X}{N} \tag{1.28}$$

be the average of all of the measurements. It is clear that $\sigma_{\tilde{x}} = \sigma_X / N$ (see Exercise 1.8). Hence, we deduce from Equation (1.27) that

$$\sigma_{\tilde{x}} = \frac{\sigma_x}{\sqrt{N}}. \tag{1.29}$$

We conclude that we can reduce the uncertainty in the measured value of x by making a large number of independent measurements, and averaging the results. Unfortunately, the error scales as one over the square root of the number of measurements, which means that increasing the number of measurements by a factor of 10 only reduces the error by a factor of $\sqrt{10} = 3.16$.

Suppose that

$$z = x\,y. \tag{1.30}$$

It follows that

$$\overline{z} + \delta z = (\overline{x} + \delta x)(\overline{y} + \delta y) = \overline{x}\,\overline{y} + \overline{x}\,\delta y + \overline{y}\,\delta x + \delta x\,\delta y. \qquad (1.31)$$

Averaging the previous equation, making use of $\langle \delta x \rangle = \langle \delta y \rangle = \langle \delta x\,\delta y \rangle = 0$, we deduce that

$$\overline{z} = \overline{x}\,\overline{y}. \qquad (1.32)$$

In other words, not surprisingly, if z is the product of x and y then the mean value of z is the product of the mean values of x and y. Taking the difference between the previous two equations, we obtain

$$\delta z = \overline{x}\,\delta y + \overline{y}\,\delta x, \qquad (1.33)$$

where we have neglected the term that is second order in the errors (assuming that the errors are relatively small). It follows that

$$(\delta z)^2 = \overline{y}^2\,(\delta x)^2 + \overline{x}^2\,(\delta y)^2 + 2\,\overline{x}\,\overline{y}\,\delta x\,\delta y. \qquad (1.34)$$

Averaging the previous equation, making use of $\overline{z} = \overline{x}\,\overline{y}$, $\langle \delta x\,\delta y \rangle = 0$, and $\langle (\delta x)^2 \rangle = \sigma_x^2$, $\langle (\delta y)^2 \rangle = \sigma_y^2$, $\langle (\delta z)^2 \rangle = \sigma_z^2$, we obtain

$$\frac{\sigma_z^2}{\overline{z}^2} = \frac{\sigma_x^2}{\overline{x}^2} + \frac{\sigma_y^2}{\overline{y}^2}. \qquad (1.35)$$

This expression specifies how the errors propagate in Equation (1.30).

A straightforward generalization of the previous analysis reveals that if

$$z = \frac{x}{y} \qquad (1.36)$$

then

$$\overline{z} = \frac{\overline{x}}{\overline{y}}, \qquad (1.37)$$

$$\frac{\sigma_z^2}{\overline{z}^2} = \frac{\sigma_x^2}{\overline{x}^2} + \frac{\sigma_y^2}{\overline{y}^2}. \qquad (1.38)$$

Finally, suppose that

$$z = f(x, y), \qquad (1.39)$$

where $f(x, y)$ is a general function of two variables. We can write

$$\overline{z} + \delta z = f(\overline{x} + \delta x, \overline{y} + \delta y). \qquad (1.40)$$

Performing a Taylor expansion of the right-hand side of the previous equation, and neglecting terms that are second order, or higher, in the small errors, we obtain

$$\overline{z} + \delta z = f(\overline{x}, \overline{y}) + \frac{\partial f(\overline{x}, \overline{y})}{\partial \overline{x}}\,\delta x + \frac{\partial f(\overline{x}, \overline{y})}{\partial \overline{y}}\,\delta y, \qquad (1.41)$$

It follows that

$$\overline{z} = f(\overline{x}, \overline{y}), \qquad (1.42)$$

$$\delta z = \frac{\partial f(\overline{x}, \overline{y})}{\partial \overline{x}}\,\delta x + \frac{\partial f(\overline{x}, \overline{y})}{\partial \overline{y}}\,\delta y. \qquad (1.43)$$

Taking the square of the previous equation, and averaging, we deduce that

$$\sigma_z^2 = \left[\frac{\partial f(\overline{x}, \overline{y})}{\partial \overline{x}}\right]^2 \sigma_x^2 + \left[\frac{\partial f(\overline{x}, \overline{y})}{\partial \overline{y}}\right]^2 \sigma_y^2. \qquad (1.44)$$

The generalization to a function of more that two variables is straightforward.

1.6   EXERCISES

1.1 Farmer Jones has recently brought a 40 acre field, and wishes to replace the surrounding fence. Given that the field is square, what length of fencing (in meters) should Farmer Jones purchase? (1 acre equals 43 560 square feet.) [Ans: 1.609×10^3 m.]

1.2 How many seconds are there in a (Julian) year? [Ans: 31 557 600 s.]

1.3 Large-scale water resources in the United States (of America) are conventionally measured in acre-feet. One acre-foot is the volume of a sheet of water of uniform depth one foot that extends over an acre. How many cubic meters are in one acre-foot? [Ans: 1233.48 m^3.]

1.4 The recommended tire pressure in a 2021 Honda Civic is 32 psi (pounds per square inch). What is this pressure in atmospheres? (1 atmosphere, which is the mean pressure of the atmosphere at sea level, is 10^5 Pa.) [Ans: 2.21 atmospheres.]

1.5 The speed of sound, v, in a gas might plausibly depend on the pressure, p, the density, ρ, and the volume, V, of the gas. Use dimensional analysis to determine the exponents x, y, and z in the formula

$$v = C\, p^x\, \rho^y\, V^z,$$

where C is a dimensionless constant. [Ans: $v = C\sqrt{p/\rho}$.]

1.6 The propagation speed of a water wave, v, in shallow water might plausibly depend on the density of the water, ρ, the depth of the water, d, and the acceleration due to gravity, g. Use dimensional analysis to determine the exponents x, y, and z in the formula

$$v = C\, \rho^x\, g^y\, d^z,$$

where C is a dimensionless constant. [Ans: $v = C\sqrt{g\,d}$.]

1.7 The additional (to atmospheric pressure) pressure a depth h below sea level might plausibly depend on the depth, the density of sea water, ρ, and the acceleration due to gravity, g. Use dimensional analysis to determine the exponents x, y, and z in the formula

$$p = C\, \rho^x\, g^y\, h^z,$$

where C is a dimensionless constant. [Ans: $p = C\, \rho g\, h$.]

1.8 Suppose that $z = a\,x + b\,y$, where x, y are physical quantities, and a, b are constants. Express σ_z in terms of a, b, σ_x, and σ_y. [Ans: $\sigma_z^2 = a^2\,\sigma_x^2 + b^2\,\sigma_y^2$.]

1.9 Suppose that $z = a\,x^\mu\,y^\nu$, where x, y are physical quantities, and a, μ, ν are constants. Express σ_z in terms of a, μ, ν, $\bar{z}$, $\bar{x}$, $\bar{y}$, σ_x, and σ_y. [Ans: $\sigma_z^2/\bar{z}^2 = \mu^2\,\sigma_x^2/\bar{x}^2 + \nu^2\,\sigma_y^2/\bar{y}^2$.]

1.10 Suppose that $z = a\,\exp(\mu\,x)$, where x is a physical quantity, and a, μ are constants. Express σ_z in terms of a, μ, $\bar{z}$, and σ_x. [Ans: $\sigma_z^2/\bar{z}^2 = \mu^2\,\sigma_x^2$.]

1.11 A measurement of the length of the side of a cube yields the result 1.237 ± 0.034 m. What is the volume of the cube? Include an estimate of the error in your answer. [Ans: 1.89 ± 0.16 m^3.]

CHAPTER 2

Motion in One Dimension

2.1 INTRODUCTION

The purpose of this chapter is to introduce the concepts of *displacement*, *velocity*, and *acceleration*. For the sake of simplicity, we shall restrict our attention to one-dimensional motion.

2.2 DISPLACEMENT

Consider a rigid body moving in one dimension. For instance, a train traveling down a straight railroad track, or a truck driving down an interstate in Kansas. Suppose that we have a team of observers who continually report the location of this body to us as time progresses. To be more exact, our observers report the distance, x, of the body from some arbitrarily chosen reference point located on the track upon which it is constrained to move. This point is known as the *origin* of our coordinate system. A positive x value implies that the body is located x meters to the right of the origin, whereas a negative x value implies that the body is located $|x|$ meters to the left of the origin. Here, x is termed the **displacement** of the body from the origin. See Figure 2.1. Of course, if the body is extended then our observers will have to report the displacement, x, of some conveniently chosen reference point on the body (e.g., its center of mass) from the origin.

Our information regarding the body's motion consists of a set of data points, each specifying the displacement, x, of the body at some time t. It is usually illuminating to graph these points. Figure 2.2 shows an example of such a graph. As is often the case, it is possible to fit the data points appearing in this graph using a relatively simple analytic

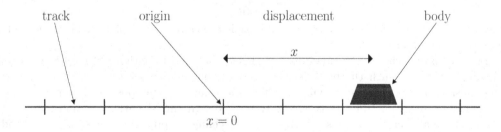

Figure 2.1 Motion in one dimension

DOI: 10.1201/9781003198642-2

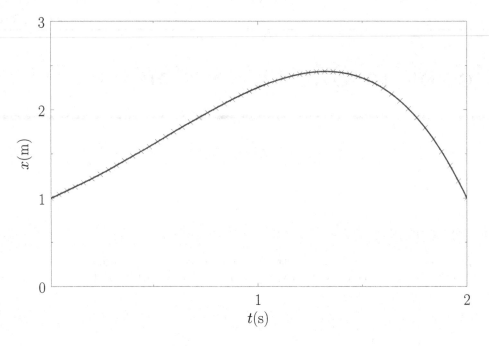

Figure 2.2 Graph of displacement versus time

curve. Indeed, the curve associated with Figure 2.2 is

$$x = 1 + t + \frac{t^2}{2} - \frac{t^4}{4}. \tag{2.1}$$

2.3 VELOCITY

Both Figure 2.2 and Equation (2.1) effectively specify the location of the body whose motion we are studying as time progresses. Let us now consider how we can use this information to determine the body's instantaneous velocity as a function of time. The conventional definition of velocity is as follows:

Velocity is the rate of change of displacement with time.

This definition implies that

$$v = \frac{\Delta x}{\Delta t}, \tag{2.2}$$

where v is the body's velocity at time t, and Δx is the change in displacement of the body between times t and $t + \Delta t$.

How should we choose the time interval Δt appearing in Equation (2.2)? Obviously, in the simple case in which the body is moving with constant velocity, we can make Δt as large or small as we like, and it will not affect the value of v. Suppose, however, that v is constantly changing in time, as is generally the case. In this situation, Δt must be kept sufficiently small that the body's velocity does not change appreciably between times t and $t + \Delta t$. If Δt is made too large then Equation (2.2) merely gives the average velocity of the body between times t and $t + \Delta t$.

Suppose that we require a general expression for instantaneous velocity that is valid irrespective of how rapidly or slowly the body's velocity changes in time. We can achieve this goal by taking the limit of Equation (2.2) as Δt approaches zero. This ensures that,

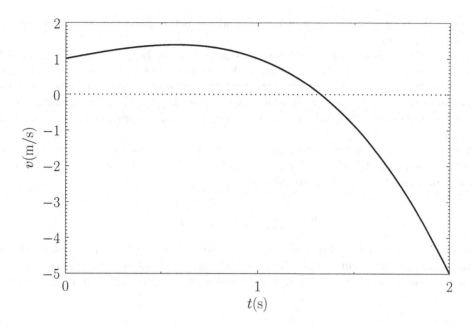

Figure 2.3 Graph of instantaneous velocity versus time associated with the motion specified in Figure 2.2

no matter how rapidly v varies with time, the velocity of the body is always approximately constant in the interval t to $t + \Delta t$. Thus,

$$v = \lim_{\Delta t \to 0} \frac{\Delta x}{\Delta t} = \frac{dx}{dt},$$ (2.3)

where dx/dt represents the derivative of x with respect to t. The previous definition is particularly useful if we can represent $x(t)$ as an analytic function, because it allows us to immediately evaluate the instantaneous velocity, $v(t)$, via the rules of calculus. Thus, if $x(t)$ is given by Equation (2.1) then

$$v \equiv \frac{dx}{dt} = 1 + t - t^3.$$ (2.4)

Figure 2.3 shows the graph of v versus t obtained from the previous expression. Note that when v is positive the body is moving to the right (i.e., x is increasing in time). Likewise, when v is negative the body is moving to the left (i.e., x is decreasing in time). Finally, when $v = 0$ the body is instantaneously at rest.

The terms velocity and speed are often confused with one another. A velocity can be either positive or negative, depending on the direction of motion. The conventional definition of *speed* is that it is the magnitude of velocity (i.e., it is v with the sign stripped off). It follows that a body can never possess a negative speed.

2.4 ACCELERATION

The conventional definition of acceleration is as follows:

Acceleration is the rate of change of velocity with time.

This definition implies that

$$a = \frac{\Delta v}{\Delta t},$$ (2.5)

where a is the body's acceleration at time t, and Δv is the change in velocity of the body between times t and $t + \Delta t$.

How should we choose the time interval Δt appearing in Equation (2.5)? Again, in the simple case in which the body is moving with constant acceleration, we can make Δt as large or small as we like, and it will not affect the value of a. Suppose, however, that a is constantly changing in time, as is generally the case. In this situation, Δt must be kept sufficiently small that the body's acceleration does not change appreciably between times t and $t + \Delta t$.

A general expression for instantaneous acceleration that is valid irrespective of how rapidly or slowly the body's acceleration changes in time can be obtained by taking the limit of Equation (2.5) as Δt approaches zero:

$$a = \lim_{\Delta t \to 0} \frac{\Delta v}{\Delta t} = \frac{dv}{dt} = \frac{d^2 x}{dt^2}, \tag{2.6}$$

where use has been made of Equation (2.3). The previous definition is particularly useful if we can represent $x(t)$ as an analytic function, because it allows us to immediately evaluate the instantaneous acceleration, $a(t)$, via the rules of calculus. Thus, if $x(t)$ is given by Equation (2.1) then

$$a \equiv \frac{d^2 x}{dt^2} = 1 - 3 t^2. \tag{2.7}$$

Figure 2.4 shows the graph of a versus time obtained from the previous expression. Note that when a is positive the body is accelerating to the right (i.e., v is increasing in time). Likewise, when a is negative the body is decelerating (i.e., v is decreasing in time).

Fortunately, it is generally not necessary to evaluate the rate of change of acceleration with time, because this quantity does not appear in Newton's laws of motion.

2.5   MOTION WITH CONSTANT VELOCITY

The simplest type of motion (excluding the trivial case in which the body under investigation remains at rest) consists of motion with constant velocity. This type of motion occurs in everyday life whenever an object slides over a horizontal, low-friction surface. For instance, when a puck slides across a hockey rink.

Figure 2.5 shows the graph of displacement versus time for a body moving with a constant velocity. It can be seen that the graph consists of a straight-line. This line can be represented algebraically as

$$x = x_0 + v t. \tag{2.8}$$

Here, x_0 is the displacement at time $t = 0$. This quantity can be determined from the graph as the intercept of the straight-line with the x-axis. Likewise, $v = dx/dt$ is the constant velocity of the body. This quantity can be determined from the graph as the gradient of the straight-line (i.e., the ratio $\Delta x/\Delta t$, as shown). Note that $a \equiv d^2 x/dt^2 = 0$, as expected.

Figure 2.6 shows a displacement versus time graph for a slightly more complicated case of motion with constant velocity. The body in question moves to the right (because x is clearly increasing with t) with a constant velocity (because the graph is a straight-line) between times A and B. The body then moves to the right (because x is still increasing in time) with a somewhat larger constant velocity (because the graph is again a straight-line, but possesses a larger gradient than before) between times B and C. The body remains at rest (because the graph is horizontal) between times C and D. Finally, the body moves to the left (because x is decreasing with t) with a constant velocity (because the graph is a straight-line) between times D and E.

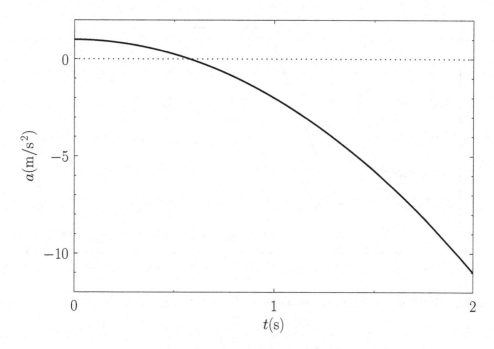

Figure 2.4 Graph of instantaneous acceleration versus time associated with the motion specified in Figure 2.2.

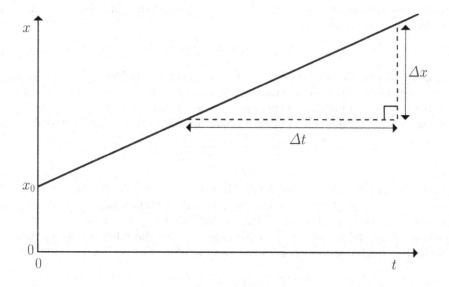

Figure 2.5 Graph of displacement versus time for a body moving with constant velocity.

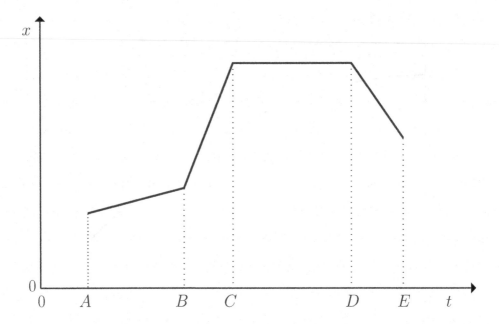

Figure 2.6 Graph of displacement versus time

2.6   MOTION WITH CONSTANT ACCELERATION

Motion with constant acceleration occurs in everyday life whenever an object is dropped. The object moves downward with the constant acceleration $9.81\,\mathrm{m\,s^{-2}}$, under the influence of gravity.

Figure 2.7 shows the graphs of displacement versus time and velocity versus time for a body moving with constant acceleration. It can be seen that the displacement-time graph consists of a curved-line whose gradient (slope) is increasing in time. This line can be represented algebraically as

$$x = x_0 + v_0\,t + \frac{1}{2}\,a\,t^2. \qquad (2.9)$$

Here, x_0 is the displacement at time $t = 0$. This quantity can be determined from the graph as the intercept of the curved-line with the x-axis. Likewise, v_0 is the body's instantaneous velocity at time $t = 0$. Finally, a is the body's constant acceleration.

The velocity-time graph consists of a straight-line that can be represented algebraically as

$$v \equiv \frac{dx}{dt} = v_0 + a\,t. \qquad (2.10)$$

The quantity v_0 is determined from the graph as the intercept of the straight-line with the x-axis. The constant acceleration, a, can be determined graphically as the gradient of the straight-line (i.e., the ratio $\Delta v/\Delta t$, as shown). Note that $dv/dt = a$, as expected.

Equations (2.9) and (2.10) can be rearranged to give the following set of three useful formulae that characterize motion with constant acceleration:

$$s = v_0\,t + \frac{1}{2}\,a\,t^2, \qquad (2.11)$$

$$v = v_0 + a\,t, \qquad (2.12)$$

$$v^2 = v_0^2 + 2\,a\,s. \qquad (2.13)$$

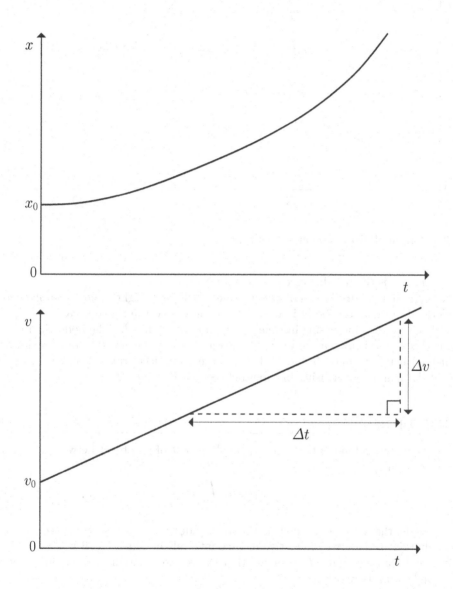

Figure 2.7 Graphs of displacement versus time (Top) and velocity versus time (Bottom) for a body moving with constant acceleration

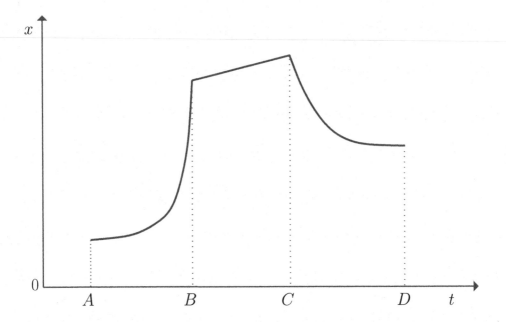

Figure 2.8 Graph of displacement versus time

Here, $s = x - x_0$ is the net distance traveled after t seconds.

Figure 2.8 shows a displacement versus time graph for a slightly more complicated case of accelerated motion. The body in question accelerates to the right [because the gradient (slope) of the graph is increasing in time] between times A and B. The body then moves to the right (because x is increasing in time) with a constant velocity (because the graph is a straight-line) between times B and C. Finally, the body decelerates [because the gradient (slope) of the graph is decreasing in time] between times C and D.

2.7 USEFUL RESULTS

Because $v = dx/dt$, it follows that the net displacement of an object between times t_1 and t_2 can be written

$$x(t_2) - x(t_1) = \int_{t_1}^{t_2} v(t') \, dt'. \tag{2.14}$$

In other words, the net displacement is the area under the object's $v(t)$ curve. (Note that the contribution to the area is negative in time intervals in which $v < 0$.)

Likewise, because $a = dv/dt$, it follows that the velocity increment of an object between times t_1 and t_2 can be written

$$v(t_2) - v(t_1) = \int_{t_1}^{t_2} a(t') \, dt'. \tag{2.15}$$

In other words, the velocity increment is the area under the object's $a(t)$ curve. (Again, the contribution to the area is negative in time intervals in which $a < 0$.)

2.8 FREE-FALL UNDER GRAVITY

Galileo was the first scientist to appreciate that, neglecting the effect of air resistance, all bodies in free-fall close to the Earth's surface accelerate vertically downward with the same acceleration; namely, $g = 9.81\,\mathrm{m\,s^{-2}}$.[1] The neglect of air resistance is a fairly good approximation for large objects that travel relatively slowly (e.g., a shot-put, or a basketball), but becomes a poor approximation for small objects that travel relatively rapidly (e.g., a golf-ball, or a bullet fired from a pistol). (See Section 4.9.)

Equations (2.11)–(2.13) can easily be modified to deal with the special case of an object free-falling under gravity:

$$s = v_0\,t - \frac{1}{2}\,g\,t^2,\qquad(2.16)$$

$$v = v_0 - g\,t,\qquad(2.17)$$

$$v^2 = v_0^2 - 2\,g\,s.\qquad(2.18)$$

Here, $g = 9.81\,\mathrm{m\,s^{-2}}$ is the downward acceleration due to gravity, s is the distance the object has moved vertically between times $t = 0$ and t (if $s > 0$ then the object has risen s meters, else if $s < 0$ then the object has fallen $|s|$ meters), and v_0 is the object's instantaneous velocity at $t = 0$. Finally, v is the object's instantaneous velocity at time t.

Let us illustrate the use of Equations (2.16)–(2.18). Suppose that a ball is released from rest, and allowed to fall under the influence of gravity. How long does it take the ball to fall h meters? According to Equation (2.16) [with $v_0 = 0$ (because the ball is released from rest), and $s = -h$ (because we wish the ball to fall h meters)], $h = g\,t^2/2$, so the time of fall is

$$t = \sqrt{\frac{2\,h}{g}}.\qquad(2.19)$$

What is the final speed, u, of the ball? According to Equation (2.18) [with $v_0 = 0$, $s = -h$, and $|v| = u$],

$$u = \sqrt{2\,g\,h}.\qquad(2.20)$$

Suppose that a ball is thrown vertically upward from ground level with velocity u. To what height does the ball rise, how long does it remain in the air, and with what velocity does it strike the ground? The ball attains its maximum height when it is momentarily at rest (i.e., when $v = 0$). According to Equation (2.17) (with $v_0 = u$), this occurs at time $t = u/g$. It follows from Equation (2.16) (with $v_0 = u$, $t = u/g$, and $s = h$) that the maximum height of the ball is given by

$$h = \frac{u^2}{2\,g}.\qquad(2.21)$$

When the ball strikes the ground it has traveled zero net meters vertically, so $s = 0$. It follows from Equations (2.17) and (2.18) (with $v_0 = u$ and $t > 0$) that $v = -u$. In other words, the ball hits the ground with an equal and opposite velocity to that with which it was thrown into the air. Because the ascent and decent phases of the ball's trajectory are

---

[1] Actually, the acceleration due to gravity varies slightly over the Earth's surface because of the combined effects of the Earth's rotation and the Earth's slightly flattened shape. See Sections 12.3 and 15.6. The acceleration at the poles is about $9.834\,\mathrm{m\,s^{-2}}$, whereas the acceleration at the equator is only $9.780\,\mathrm{m\,s^{-2}}$. The average acceleration over the Earth's surface is $9.81\,\mathrm{m\,s^{-2}}$.

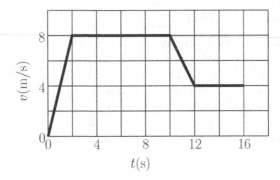

Figure 2.9 Figure for Exercise 2.1

clearly symmetric, the ball's time of flight is simply twice the time required for the ball to attain its maximum height; that is,

$$t = \frac{2u}{g}.\qquad(2.22)$$

2.9 EXERCISES

2.1 Consider the motion of the object whose velocity-time graph is given in Figure 2.9.

  (a) What is the acceleration of the object between times $t = 0$ and $t = 2$? [Ans: $4\,\mathrm{m\,s^{-2}}$.]

  (b) What is the acceleration of the object between times $t = 10$ and $t = 12$? [Ans: $-2\,\mathrm{m\,s^{-2}}$.]

  (c) What is the net displacement of the object between times $t = 0$ and $t = 16$? [Ans: 100 m.]

2.2 Consider the motion of the object whose velocity-time graph is given in Figure 2.10. What is the net displacement of the object between times $t = 0$ and $t = 16$? [Ans: 8 m.]

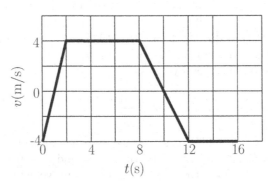

Figure 2.10 Figure for Exercise 2.2.

2.3 The velocity of a certain body as a function of time is given by

$$v = t^2 - 3t + 2,$$

where v is measured in meters per second and t is measured in seconds.

(a) At which two times is the body instantaneously at rest? [Ans: $t = 1\,$s and $t = 2\,$s.]

(b) What is the body's minimum velocity? [Ans: $-0.25\,$m/s.]

(c) What is the body's net displacement between the two times at which it is instantaneously at rest? [Ans: $-0.167\,$m.]

2.4 In a speed trap, two pressure-activated strips are placed $120\,$m apart on a highway on which the speed limit is $85\,$km/h. A driver traveling at $110\,$km/h notices a police car just as he/she activates the first strip, and immediately slows down. What constant deceleration is needed to ensure that the car's average speed is within the speed limit when the car crosses the second strip? [Ans: $2.73\,\mathrm{m\,s^{-2}}$.]

2.5 Let $v(t)$ represent the velocity of a car versus time. Suppose that $\bar{v}$ is the car's average velocity between times t_1 and t_2.

(a) Demonstrate that
$$\int_{t_1}^{t_2} [v(t') - \bar{v}]\, dt' = 0.$$

(b) In some countries, you can be prosecuted for speeding if your car's average speed between two checkpoints (which may be miles apart) exceeds the posted speed limit. Demonstrate that if your car's average speed really did exceed the speed limit then it is impossible for its instantaneous speed not to have exceeded the limit at some point between the two checkpoints.

2.6 In 1886, Steve Brodie achieved notoriety by jumping off the recently completed Brooklyn Bridge, for a bet, and surviving. Given that the bridge rises $135\,$ft over the East River, how long would Mr. Brodie have been in the air, and with what speed would he have struck the water? You may neglect air resistance (which is not particularly realistic, in this case). [Ans: $2.896\,$s, and $28.41\,\mathrm{m\,s^{-1}}$ (i.e., $63.6\,$mi/h)!]

2.7 With what speed, in miles per hour, would you need to throw a ball vertically upward from ground level in order for it to attain a maximum height above the ground of 1 mile? Neglect air resistance (which is not at all realistic, in this case). [Ans: $397.5\,$mph.]

2.8 A boat takes a time t_1 to travel a distance a up a river, and a time t_2 to return. Show that the speed of the boat relative to the river is $a\,(t_1 + t_2)/(2\,t_1\,t_2)$. [From Lamb 1942.]

2.9 A particle is projected upward from ground level, attains a height h after t_1 seconds, and again after t_2 seconds.

(a) Show that $h = (1/2)\,g\,t_1\,t_2$.

(b) Show that the initial velocity was $(1/2)\,g\,(t_1 + t_2)$.

[From Lamb 1942.]

2.10 The speed of a car increases at a constant rate α from 0 to u, then remains constant for an interval, and finally decreases to 0 at the constant rate β. Let l be the total distance traveled.

(a) Show that the total travel time is

$$\frac{l}{u} + \frac{u}{2}\left(\frac{1}{\alpha} + \frac{1}{\beta}\right).$$

(b) For what value of u is the time least? [Ans: $\sqrt{2\,l\,\alpha\,\beta/(\alpha + \beta)}$.]

(c) What is the least travel time? [Ans: $\sqrt{2\,l\,(\alpha + \beta)/(\alpha\,\beta)}$.]

[From Lamb 1942.]

2.11 If the displacements of a point object moving with constant acceleration are x_1, x_2, and x_3 at times t_1, t_2, and t_3, respectively, show that the acceleration is

$$\frac{2\left[(x_2 - x_3)\,t_1 + (x_3 - x_1)\,t_2 + (x_1 - x_2)\,t_3\right]}{(t_2 - t_3)\,(t_3 - t_1)\,(t_1 - t_2)}.$$

[From Lamb 1942.]

2.12 If t is a quadratic function of x, show that the acceleration varies inversely as the cube of the distance from a fixed point. [From Lamb 1942.]

2.13 If x^2 is a quadratic function of t, show that the acceleration varies as $1/x^3$, except in a particular case. [From Lamb 1942.]

2.14 Prove that a point object cannot move such that its velocity varies as the distance it has traveled from rest. Can it move such that the velocity varies as the square root of the distance? [Ans: Yes.] [From Lamb 1942.]

CHAPTER 3

Motion in Three Dimensions

3.1 INTRODUCTION

The purpose of this chapter is to generalize the previously introduced concepts of displacement, velocity, and acceleration in order to deal with motion in three dimensions.

3.2 VECTOR MATHEMATICS

3.2.1 Scalars and Vectors

Consider the motion of a body moving in three dimensions. The body's instantaneous position is most conveniently specified by giving its displacement from the origin of our coordinate system. Note, however, that in three dimensions such a displacement possesses both magnitude and direction. In other words, we not only have to specify how far the body is situated from the origin, we also have to specify in which direction it lies. A quantity that possesses both magnitude and direction is termed a **vector**. By contrast, a quantity that possesses only magnitude is termed a **scalar**. (A more exact definition of vector and scalar quantities is given in Section 3.2.4.) Mass and time are scalar quantities. However, in general, displacement is a vector. In fact, vector displacement is the prototype of all vectors.

A general vector obeys the same algebra as a displacement in space, and may thus be represented geometrically by a straight-line, $\overrightarrow{PQ}$ (say), where the arrow indicates the direction of the displacement (i.e., from point P to point Q). See Figure 3.1. The magnitude of the vector is represented by the length of the straight-line.

It is conventional to denote vectors by bold-faced symbols (e.g., $\mathbf{a}$, $\mathbf{F}$) and scalars by non-bold-faced symbols (e.g., r, S). The magnitude of a general vector, $\mathbf{a}$, is denoted $|\mathbf{a}|$, or just a. It is convenient to define a vector with zero magnitude; this is denoted $\mathbf{0}$, and has

Figure 3.1 A vector. (Reproduced from Fitzpatrick 2008. Courtesy of Jones & Bartlett.)

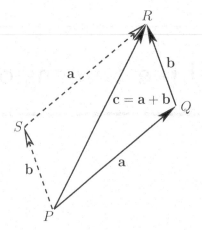

Figure 3.2 Vector addition. (Reproduced from Fitzpatrick 2008. Courtesy of Jones & Bartlett.)

no direction. Finally, two vectors, **a** and **b**, are said to be equal when their magnitudes and directions are both identical.

3.2.2 Vector Algebra

Suppose that the displacements $\overrightarrow{PQ}$ and $\overrightarrow{QR}$, shown in Figure 3.2, represent the vectors **a** and **b**, respectively. It can be seen that the result of combining these two displacements is to give the net displacement $\overrightarrow{PR}$. Hence, if $\overrightarrow{PR}$ represents the vector **c** then we can write

$$\mathbf{c} = \mathbf{a} + \mathbf{b}. \tag{3.1}$$

This defines *vector addition*. By completing the parallelogram $PQRS$, we can also see that

$$\overrightarrow{PR} = \overrightarrow{PQ} + \overrightarrow{QR} = \overrightarrow{PS} + \overrightarrow{SR}. \tag{3.2}$$

However, $\overrightarrow{PS}$ has the same length and direction as $\overrightarrow{QR}$, and, thus, represents the same vector, **b**. Likewise, $\overrightarrow{PQ}$ and $\overrightarrow{SR}$ both represent the vector **a**. Thus, the previous equation is equivalent to

$$\mathbf{c} = \mathbf{a} + \mathbf{b} = \mathbf{b} + \mathbf{a}. \tag{3.3}$$

We conclude that the addition of vectors is *commutative*. It can also be shown that the *associative* law holds; that is,

$$\mathbf{a} + (\mathbf{b} + \mathbf{c}) = (\mathbf{a} + \mathbf{b}) + \mathbf{c}. \tag{3.4}$$

The null vector, **0**, is represented by a displacement of zero length and arbitrary direction. Because the result of combining such a displacement with a finite-length displacement is the same as the latter displacement by itself, it follows that

$$\mathbf{a} + \mathbf{0} = \mathbf{a}, \tag{3.5}$$

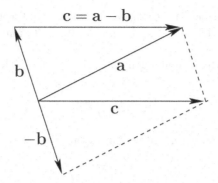

Figure 3.3   Vector subtraction. (Reproduced from Fitzpatrick 2008. Courtesy of Jones & Bartlett.)

where $\mathbf{a}$ is a general vector. The negative of $\mathbf{a}$ is defined as that vector which has the same magnitude, but acts in the opposite direction, and is denoted $-\mathbf{a}$. The sum of $\mathbf{a}$ and $-\mathbf{a}$ is thus the null vector; that is,

$$\mathbf{a} + (-\mathbf{a}) = \mathbf{0}. \tag{3.6}$$

We can also define the difference of two vectors, $\mathbf{a}$ and $\mathbf{b}$, as

$$\mathbf{c} = \mathbf{a} - \mathbf{b} = \mathbf{a} + (-\mathbf{b}). \tag{3.7}$$

This definition of *vector subtraction* is illustrated in Figure 3.3.

If $n > 0$ is a scalar then the expression $n\,\mathbf{a}$ denotes a vector whose direction is the same as $\mathbf{a}$, and whose magnitude is n times that of $\mathbf{a}$. (This definition becomes obvious when n is an integer.) If n is negative then, because $n\,\mathbf{a} = |n|\,(-\mathbf{a})$, it follows that $n\,\mathbf{a}$ is a vector whose magnitude is $|n|$ times that of $\mathbf{a}$, and whose direction is opposite to $\mathbf{a}$. These definitions imply that if n and m are two scalars then

$$n\,(m\,\mathbf{a}) = n\,m\,\mathbf{a} = m\,(n\,\mathbf{a}), \tag{3.8}$$

$$(n + m)\,\mathbf{a} = n\,\mathbf{a} + m\,\mathbf{a}, \tag{3.9}$$

$$n\,(\mathbf{a} + \mathbf{b}) = n\,\mathbf{a} + n\,\mathbf{b}. \tag{3.10}$$

3.2.3   Cartesian Components of a Vector

Consider a Cartesian coordinate system $Oxyz$, consisting of an origin, O, and three mutually perpendicular coordinate axes, Ox, Oy, and Oz. See Figure 3.4. Such a system is said to be *right-handed* if, when looking along the Oz direction, a $90°$ clockwise rotation about Oz is required to take Ox into Oy. Otherwise, it is said to be left-handed. It is conventional to always use a right-handed coordinate system.

It is convenient to define unit vectors, $\mathbf{e}_x$, $\mathbf{e}_y$, and $\mathbf{e}_z$, parallel to Ox, Oy, and Oz, respectively. Incidentally, a unit vector is a vector whose magnitude is unity. The position vector, $\mathbf{r}$, of some general point P whose Cartesian coordinates are (x, y, z) is then given by

$$\mathbf{r} = x\,\mathbf{e}_z + y\,\mathbf{e}_y + z\,\mathbf{e}_z. \tag{3.11}$$

In other words, we can get from O to P by moving a distance x parallel to Ox, then a distance y parallel to Oy, and then a distance z parallel to Oz. Similarly, if $\mathbf{a}$ is an arbitrary vector then

$$\mathbf{a} = a_x\,\mathbf{e}_x + a_y\,\mathbf{e}_y + a_z\,\mathbf{e}_z, \tag{3.12}$$

where a_x, a_y, and a_z are termed the *Cartesian components* of **a**. It is conventional to write $\mathbf{a} \equiv (a_x, a_y, a_z)$. It follows that $\mathbf{e}_x \equiv (1, 0, 0)$, $\mathbf{e}_y \equiv (0, 1, 0)$, and $\mathbf{e}_z \equiv (0, 0, 1)$. Of course, $\mathbf{0} \equiv (0, 0, 0)$.

According to the three-dimensional generalization of the Pythagorean theorem, the distance $OP \equiv |\mathbf{r}| = r$ is given by

$$r = \sqrt{x^2 + y^2 + z^2}. \tag{3.13}$$

By analogy, the magnitude of a general vector, **a**, takes the form

$$a = \sqrt{a_x^2 + a_y^2 + a_z^2}. \tag{3.14}$$

If $\mathbf{a} \equiv (a_x, a_y, a_z)$ and $\mathbf{b} \equiv (b_x, b_y, b_z)$ then it is easily demonstrated that

$$\mathbf{a} + \mathbf{b} \equiv (a_x + b_x, a_y + b_y, a_z + b_z). \tag{3.15}$$

Furthermore, if n is a scalar then it is apparent that

$$n\,\mathbf{a} \equiv (n\,a_x, n\,a_y, n\,a_z). \tag{3.16}$$

3.2.4 Coordinate Transformations

A Cartesian coordinate system allows position and direction in space to be represented in a very convenient manner. Unfortunately, such a coordinate system also introduces arbitrary elements into our analysis. After all, two independent observers might well choose Cartesian coordinate systems with different origins, and different orientations of the coordinate axes. In general, a given vector, **a**, will have different sets of components in these two coordinate systems. However, the direction and magnitude of **a** are the same in both cases. Hence, the two sets of components must be related to one another in a very particular fashion. Actually, because vectors are represented by moveable line elements in space (i.e., in Figure 3.2, $\overrightarrow{PQ}$ and $\overrightarrow{SR}$ represent the same vector), it follows that the components of a general vector are not affected by a simple shift in the origin of a Cartesian coordinate system. On the other hand, the components are modified when the coordinate axes are rotated.

Suppose that we transform to a new coordinate system, $Ox'y'z'$, which has the same origin as $Oxyz$, and is obtained by rotating the coordinate axes of $Oxyz$ through an angle

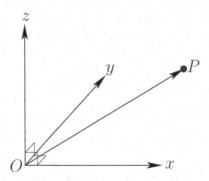

Figure 3.4 A right-handed Cartesian coordinate system. (Reproduced from Fitzpatrick 2008. Courtesy of Jones & Bartlett.)

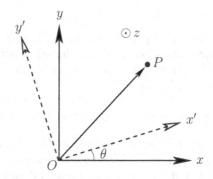

Figure 3.5 Rotation of the coordinate axes about Oz. (Reproduced from Fitzpatrick 2008. Courtesy of Jones & Bartlett.)

θ about Oz. See Figure 3.5. Let the coordinates of a general point P be (x, y, z) in $Oxyz$ and (x', y', z') in $Ox'y'z'$. According to simple trigonometry, these two sets of coordinates are related to one another via the transformation

$$x' = x \cos \theta + y \sin \theta, \tag{3.17}$$

$$y' = -x \sin \theta + y \cos \theta, \tag{3.18}$$

$$z' = z. \tag{3.19}$$

Consider the vector displacement $\mathbf{r} \equiv \overrightarrow{OP}$. Note that this displacement is represented by the same symbol, $\mathbf{r}$, in both coordinate systems, because the magnitude and direction of $\mathbf{r}$ are manifestly independent of the orientation of the coordinate axes. The coordinates of $\mathbf{r}$ do depend on the orientation of the axes; that is, $\mathbf{r} \equiv (x, y, z)$ in $Oxyz$, and $\mathbf{r} \equiv (x', y', z')$ in $Ox'y'z'$. However, they must depend in a very specific manner [i.e., Equations (3.17)–(3.19)] that preserves the magnitude and direction of $\mathbf{r}$.

The components of a general vector, $\mathbf{a}$, transform in an analogous manner to Equations (3.17)–(3.19); that is,

$$a_{x'} = a_x \cos \theta + a_y \sin \theta, \tag{3.20}$$

$$a_{y'} = -a_x \sin \theta + a_y \cos \theta, \tag{3.21}$$

$$a_{z'} = a_z. \tag{3.22}$$

Moreover, there are similar transformation rules for rotation about Ox and Oy. Equations (3.20)–(3.22) constitute the true definition of a vector. In other words, the three quantities (a_x, a_y, a_z) are the components of a vector provided that they transform under rotation of the coordinate axes about Oz in accordance with Equations (3.20)–(3.22). (And also transform correctly under rotation about Ox and Oy.) Conversely, (a_x, a_y, a_z) cannot be the components of a vector if they do not transform in accordance with Equations (3.20)–(3.22). The true definition of a scalar quantity is that it is invariant under rotation of the coordinate axes. Thus, the individual components of a vector $(a_x$, say) are real numbers, but they are not scalars. Displacement vectors, and all vectors derived from displacements (e.g., velocity and acceleration), automatically satisfy Equations (3.20)–(3.22).

3.2.5 Scalar Product

A scalar quantity is invariant under all possible rotational transformations. The individual components of a vector are not scalars because they change under transformation. Can we form a scalar out of some combination of the components of one, or more, vectors? Suppose that we were to define the "percent" product,

$$\mathbf{a} \% \mathbf{b} \equiv a_x\, b_z + a_y\, b_x + a_z\, b_y = \text{scalar number}, \tag{3.23}$$

for general vectors $\mathbf{a}$ and $\mathbf{b}$. Is $\mathbf{a} \% \mathbf{b}$ invariant under transformation, as must be the case if it is a scalar number? Let us consider an example. Suppose that $\mathbf{a} \equiv (0,\,1,\,0)$ and $\mathbf{b} \equiv (1,\,0,\,0)$. It is easily seen that $\mathbf{a} \% \mathbf{b} = 1$. Let us now rotate the coordinate axes through $45°$ about Oz. In the new coordinate system, $\mathbf{a} \equiv (1/\sqrt{2},\, 1/\sqrt{2},\, 0)$ and $\mathbf{b} \equiv (1/\sqrt{2},\, -1/\sqrt{2},\, 0)$, giving $\mathbf{a} \% \mathbf{b} = 1/2$. Clearly, $\mathbf{a} \% \mathbf{b}$ is not invariant under rotational transformation, so the previous definition is a bad one.

Consider, now, the *scalar product*:

$$\mathbf{a} \cdot \mathbf{b} \equiv a_x\, b_x + a_y\, b_y + a_z\, b_z = \text{scalar number}. \tag{3.24}$$

Let us rotate the coordinate axes though θ degrees about Oz. According to Equations (3.20)–(3.22), $\mathbf{a} \cdot \mathbf{b}$ takes the form

$$\begin{aligned} \mathbf{a} \cdot \mathbf{b} &= (a_x\, \cos\theta + a_y\, \sin\theta)\,(b_x\, \cos\theta + b_y\, \sin\theta) \\ &\quad + (-a_x\, \sin\theta + a_y\, \cos\theta)\,(-b_x\, \sin\theta + b_y\, \cos\theta) + a_z\, b_z \\ &= a_x\, b_x + a_y\, b_y + a_z\, b_z \end{aligned} \tag{3.25}$$

in the new coordinate system. Thus, $\mathbf{a} \cdot \mathbf{b}$ is invariant under rotation about Oz. It is easily demonstrated that it is also invariant under rotation about Ox and Oy. We conclude that $\mathbf{a} \cdot \mathbf{b}$ is a true scalar, and that the definition (3.24) is a good one. Incidentally, $\mathbf{a} \cdot \mathbf{b}$ is the only simple combination of the components of two vectors that transforms like a scalar. It is readily shown that the scalar product is commutative and distributive: that is,

$$\mathbf{a} \cdot \mathbf{b} = \mathbf{b} \cdot \mathbf{a},$$
$$\mathbf{a} \cdot (\mathbf{b} + \mathbf{c}) = \mathbf{a} \cdot \mathbf{b} + \mathbf{a} \cdot \mathbf{c}. \tag{3.26}$$

The associative property is meaningless for the scalar product, because we cannot have $(\mathbf{a} \cdot \mathbf{b}) \cdot \mathbf{c}$, as $\mathbf{a} \cdot \mathbf{b}$ is scalar.

We have shown that the scalar product $\mathbf{a} \cdot \mathbf{b}$ is coordinate independent. But what is the geometric significance of this property? In the special case where $\mathbf{a} = \mathbf{b}$, we get

$$\mathbf{a} \cdot \mathbf{b} = a_x^2 + a_y^2 + a_z^2 = |\mathbf{a}|^2 = a^2. \tag{3.27}$$

So, the invariance of $\mathbf{a} \cdot \mathbf{a}$ is equivalent to the invariance of the magnitude of vector $\mathbf{a}$ under transformation.

Let us now investigate the general case. The length squared of AB in the vector triangle shown in Figure 3.6 is

$$(\mathbf{b} - \mathbf{a}) \cdot (\mathbf{b} - \mathbf{a}) = |\mathbf{a}|^2 + |\mathbf{b}|^2 - 2\,\mathbf{a} \cdot \mathbf{b}. \tag{3.28}$$

However, according to the "cosine rule" of trigonometry,

$$(AB)^2 = (OA)^2 + (OB)^2 - 2\,(OA)\,(OB)\,\cos\theta, \tag{3.29}$$

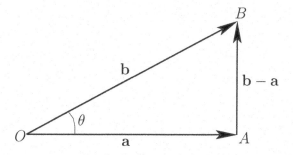

Figure 3.6 A vector triangle. (Reproduced from Fitzpatrick 2008. Courtesy of Jones & Bartlett.)

where (AB) denotes the length of side AB, etcetera. It follows that

$$\mathbf{a} \cdot \mathbf{b} = |\mathbf{a}|\,|\mathbf{b}|\,\cos\theta. \tag{3.30}$$

In this case, the invariance of $\mathbf{a} \cdot \mathbf{b}$ under transformation is equivalent to the invariance of the angle subtended between the two vectors. Note that if $\mathbf{a} \cdot \mathbf{b} = 0$ then either $|\mathbf{a}| = 0$, $|\mathbf{b}| = 0$, or the vectors $\mathbf{a}$ and $\mathbf{b}$ are mutually perpendicular. The angle subtended between two vectors can easily be obtained from the scalar product:

$$\cos\theta = \frac{\mathbf{a} \cdot \mathbf{b}}{|\mathbf{a}|\,|\mathbf{b}|}. \tag{3.31}$$

3.2.6 Vector Product

We have discovered how to construct a scalar from the components of two general vectors, $\mathbf{a}$ and $\mathbf{b}$. Can we also construct a vector that is not just a linear combination of $\mathbf{a}$ and $\mathbf{b}$? Consider the following definition:

$$\mathbf{a} * \mathbf{b} \equiv (a_x\,b_x,\ a_y\,b_y,\ a_z\,b_z). \tag{3.32}$$

Is $\mathbf{a} * \mathbf{b}$ a proper vector? Suppose that $\mathbf{a} = (0,\,1,\,0)$, $\mathbf{b} = (1,\,0,\,0)$. In this case, $\mathbf{a} * \mathbf{b} = \mathbf{0}$. However, if we rotate the coordinate axes through $45°$ about Oz then $\mathbf{a} = (1/\sqrt{2},\,1/\sqrt{2},\,0)$, $\mathbf{b} = (1/\sqrt{2},\,-1/\sqrt{2},\,0)$, and $\mathbf{a} * \mathbf{b} = (1/2,\,-1/2,\,0)$. Thus, $\mathbf{a} * \mathbf{b}$ does not transform like a vector, because its magnitude depends on the choice of axes. So, previous definition is a bad one.

Consider, now, the *vector product*:

$$\mathbf{a} \times \mathbf{b} \equiv (a_y\,b_z - a_z\,b_y,\ a_z\,b_x - a_x\,b_z,\ a_x\,b_y - a_y\,b_x) = \mathbf{c}. \tag{3.33}$$

Does this rather unlikely combination transform like a vector? Let us try rotating the coordinate axes through an angle θ about Oz using Equations (3.20)–(3.22). In the new coordinate system,

$$
\begin{aligned}
c_{x'} &= (-a_x\,\sin\theta + a_y\,\cos\theta)\,b_z - a_z\,(-b_x\,\sin\theta + b_y\,\cos\theta) \\
&= (a_y\,b_z - a_z\,b_y)\,\cos\theta + (a_z\,b_x - a_x\,b_z)\,\sin\theta \\
&= c_x\,\cos\theta + c_y\,\sin\theta. \tag{3.34}
\end{aligned}
$$

Thus, the x-component of $\mathbf{a} \times \mathbf{b}$ transforms correctly. It can easily be shown that the other components transform correctly as well, and that all components also transform correctly

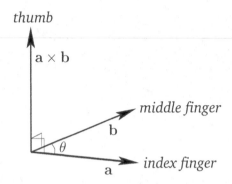

Figure 3.7 The right-hand rule for vector products. Here, θ is less than $180°$. (Reproduced from Fitzpatrick 2008. Courtesy of Jones & Bartlett.)

under rotation about Ox and Oy. Thus, $\mathbf{a} \times \mathbf{b}$ is a proper vector. Incidentally, $\mathbf{a} \times \mathbf{b}$ is the only simple combination of the components of two vectors that transforms like a vector (which is non-coplanar with $\mathbf{a}$ and $\mathbf{b}$). The vector product is *anti-commutative*,

$$\mathbf{a} \times \mathbf{b} = -\mathbf{b} \times \mathbf{a}, \tag{3.35}$$

distributive,

$$\mathbf{a} \times (\mathbf{b} + \mathbf{c}) = \mathbf{a} \times \mathbf{b} + \mathbf{a} \times \mathbf{c}, \tag{3.36}$$

but is not associative,

$$\mathbf{a} \times (\mathbf{b} \times \mathbf{c}) \neq (\mathbf{a} \times \mathbf{b}) \times \mathbf{c}. \tag{3.37}$$

The vector product transforms like a vector, which means that it must have a well-defined direction and magnitude. We can show that $\mathbf{a} \times \mathbf{b}$ is perpendicular to both $\mathbf{a}$ and $\mathbf{b}$. Consider $\mathbf{a} \cdot \mathbf{a} \times \mathbf{b}$. If this is zero then the vector product must be perpendicular to $\mathbf{a}$. Now,

$$\mathbf{a} \cdot \mathbf{a} \times \mathbf{b} = a_x \left(a_y\, b_z - a_z\, b_y \right) + a_y \left(a_z\, b_x - a_x\, b_z \right) + a_z \left(a_x\, b_y - a_y\, b_x \right)$$
$$= 0. \tag{3.38}$$

Therefore, $\mathbf{a} \times \mathbf{b}$ is perpendicular to $\mathbf{a}$. Likewise, it can be demonstrated that $\mathbf{a} \times \mathbf{b}$ is perpendicular to $\mathbf{b}$. The vectors $\mathbf{a}$, $\mathbf{b}$, and $\mathbf{a} \times \mathbf{b}$ form a right-handed set, like the unit vectors $\mathbf{e}_x$, $\mathbf{e}_y$, and $\mathbf{e}_z$. In fact, $\mathbf{e}_x \times \mathbf{e}_y = \mathbf{e}_z$. This defines a unique direction for $\mathbf{a} \times \mathbf{b}$, which is obtained from a right-hand rule. See Figure 3.7.

Let us now evaluate the magnitude of $\mathbf{a} \times \mathbf{b}$. We have

$$(\mathbf{a} \times \mathbf{b})^2 = (a_y\, b_z - a_z\, b_y)^2 + (a_z\, b_x - a_x\, b_z)^2 + (a_x\, b_y - a_y\, b_x)^2$$
$$= (a_x^2 + a_y^2 + a_z^2)(b_x^2 + b_y^2 + b_z^2) - (a_x\, b_x + a_y\, b_y + a_z\, b_z)^2$$
$$= |\mathbf{a}|^2 |\mathbf{b}|^2 - (\mathbf{a} \cdot \mathbf{b})^2$$
$$= |\mathbf{a}|^2 |\mathbf{b}|^2 - |\mathbf{a}|^2 |\mathbf{b}|^2 \cos^2 \theta = |\mathbf{a}|^2 |\mathbf{b}|^2 \sin^2 \theta. \tag{3.39}$$

Thus,

$$|\mathbf{a} \times \mathbf{b}| = |\mathbf{a}|\, |\mathbf{b}| \sin \theta, \tag{3.40}$$

where θ is the angle subtended between $\mathbf{a}$ and $\mathbf{b}$. Clearly, $\mathbf{a} \times \mathbf{a} = \mathbf{0}$ for any vector, because θ is always zero in this case. Also, if $\mathbf{a} \times \mathbf{b} = \mathbf{0}$ then either $|\mathbf{a}| = 0$, $|\mathbf{b}| = 0$, or $\mathbf{b}$ is parallel (or antiparallel) to $\mathbf{a}$.

3.3   VECTOR DISPLACEMENT, VELOCITY, AND ACCELERATION

Consider a body moving in three dimensions. Suppose that we know the Cartesian coordinates, x, y, and z, of this body as time, t, progresses. Let us consider how we can use this information to determine the body's instantaneous velocity and acceleration as functions of time.

The vector displacement of the body is given by

$$\mathbf{r}(t) = [x(t), y(t), z(t)]. \tag{3.41}$$

By analogy with the one-dimensional equation (2.3), the body's vector velocity, $\mathbf{v} = (v_x, v_y, v_z)$, is simply the derivative of $\mathbf{r}$ with respect to t. In other words,

$$\mathbf{v}(t) = \lim_{\Delta t \to 0} \frac{\mathbf{r}(t + \Delta t) - \mathbf{r}(t)}{\Delta t} = \frac{d\mathbf{r}}{dt}. \tag{3.42}$$

When written in component form, the previous definition yields

$$v_x = \frac{dx}{dt}, \tag{3.43}$$

$$v_y = \frac{dy}{dt}, \tag{3.44}$$

$$v_z = \frac{dz}{dt}. \tag{3.45}$$

Thus, the x-component of velocity is simply the time derivative of the x-coordinate, and so on.

By analogy with the one-dimensional equation (2.6), the body's vector acceleration, $\mathbf{a} = (a_x, a_y, a_z)$, is simply the derivative of $\mathbf{v}$ with respect to t. In other words,

$$\mathbf{a}(t) = \lim_{\Delta t \to 0} \frac{\mathbf{v}(t + \Delta t) - \mathbf{v}(t)}{\Delta t} = \frac{d\mathbf{v}}{dt} = \frac{d^2\mathbf{r}}{dt^2}, \tag{3.46}$$

where use has been made of Equation (3.42). When written in component form, the previous definition yields

$$a_x = \frac{dv_x}{dt} = \frac{d^2x}{dt^2}, \tag{3.47}$$

$$a_y = \frac{dv_y}{dt} = \frac{d^2y}{dt^2}, \tag{3.48}$$

$$a_z = \frac{dv_z}{dt} = \frac{d^2z}{dt^2}, \tag{3.49}$$

where use has been made of Equations (3.43)–(3.45). Thus, the x-component of acceleration is simply the time derivative of the x-component of velocity, and so on.

As an example, suppose that the coordinates of the body are given by

$$x = \sin t, \tag{3.50}$$

$$y = \cos t, \tag{3.51}$$

$$z = 3\,t. \tag{3.52}$$

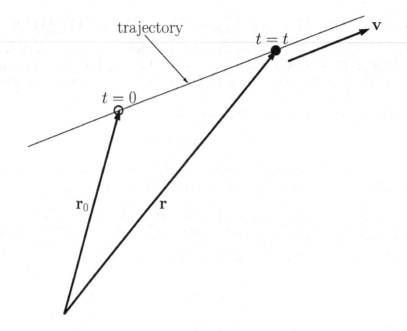

Figure 3.8 Motion with constant velocity

The corresponding components of the body's velocity are then simply

$$v_x = \frac{dx}{dt} = \cos t, \tag{3.53}$$

$$v_y = \frac{dy}{dt} = -\sin t, \tag{3.54}$$

$$v_z = \frac{dz}{dt} = 3, \tag{3.55}$$

while the components of the body's acceleration take the form

$$a_x = \frac{dv_x}{dt} = -\sin t, \tag{3.56}$$

$$a_y = \frac{dv_y}{dt} = -\cos t, \tag{3.57}$$

$$a_z = \frac{dv_z}{dt} = 0. \tag{3.58}$$

3.4   MOTION WITH CONSTANT VELOCITY

An object moving in three dimensions with constant velocity, $\mathbf{v}$, possesses a vector displacement of the form

$$\mathbf{r}(t) = \mathbf{r}_0 + \mathbf{v}\,t, \tag{3.59}$$

where the constant vector $\mathbf{r}_0$ is the displacement at time $t = 0$. Note that $d\mathbf{r}/dt = \mathbf{v}$ and $d^2\mathbf{r}/dt^2 = \mathbf{0}$, as expected. As illustrated in Figure 3.8, the object's trajectory is a straight-line that passes through point $\mathbf{r}_0$ at time $t = 0$, and runs parallel to vector $\mathbf{v}$.

3.5 MOTION WITH CONSTANT ACCELERATION

An object moving in three dimensions with constant acceleration, $\mathbf{a}$, possesses a vector displacement of the form

$$\mathbf{r}(t) = \mathbf{r}_0 + \mathbf{v}_0\, t + \frac{1}{2}\,\mathbf{a}\, t^2. \tag{3.60}$$

Hence, the object's velocity is given by

$$\mathbf{v}(t) \equiv \frac{d\mathbf{r}}{dt} = \mathbf{v}_0 + \mathbf{a}\, t. \tag{3.61}$$

Note that $d\mathbf{v}/dt = \mathbf{a}$, as expected. In the previous expressions, the constant vectors $\mathbf{r}_0$ and $\mathbf{v}_0$ are the object's displacement and velocity at time $t = 0$, respectively.

As is easily demonstrated, the vector equivalents of Equations (2.11)–(2.13) are:

$$\mathbf{s} = \mathbf{v}_0\, t + \frac{1}{2}\,\mathbf{a}\, t^2, \tag{3.62}$$

$$\mathbf{v} = \mathbf{v}_0 + \mathbf{a}\, t, \tag{3.63}$$

$$v^2 = v_0^2 + 2\,\mathbf{a} \cdot \mathbf{s}. \tag{3.64}$$

These equations fully characterize three-dimensional motion with constant acceleration. Here, $\mathbf{s} = \mathbf{r} - \mathbf{r}_0$ is the net displacement of the object between times $t = 0$ and t. Incidentally, Equation (3.64) is obtained by taking the scalar product of Equation (3.63) with itself, taking the scalar product of Equation (3.62) with $\mathbf{a}$, and then eliminating t.

3.6 PROJECTILE MOTION

As a simple illustration of the concepts introduced in the previous sections, let us examine the following well-known problem. Suppose that a projectile is launched upward from ground level, with speed v_0, making an angle θ with the horizontal. Let us determine the subsequent trajectory of the projectile, neglecting air resistance.

Our first task is to set up a suitable Cartesian coordinate system. A convenient system is illustrated in Figure 3.9. The z-axis points vertically upward (this is a standard convention), whereas the x-axis points along the projectile's initial direction of horizontal motion. Furthermore, the origin of our coordinate system corresponds to the launch point. Thus, $z = 0$ corresponds to ground level.

Neglecting air resistance, the projectile is subject to a constant acceleration $g = 9.81\,\mathrm{m\,s^{-1}}$, due to gravity, which is directed vertically downward. Thus, the projectile's vector acceleration is written

$$\mathbf{a} = (0,\, 0,\, -g). \tag{3.65}$$

Here, the minus sign indicates that the acceleration is in the minus z-direction (i.e., downward), as opposed to the plus z-direction (i.e., upward).

What is the initial vector velocity, $\mathbf{v}_0$, with which the projectile is launched into the air at (say) $t = 0$? As illustrated in Figure 3.9, given that the magnitude of this velocity is v_0, its horizontal component is directed along the x-axis, and its direction subtends an angle θ with this axis, the components of $\mathbf{v}_0$ take the form

$$\mathbf{v}_0 = (v_0 \cos\theta,\, 0,\, v_0 \sin\theta). \tag{3.66}$$

Note that $\mathbf{v}_0$ has zero component along the y-axis, which is directed into the paper in Figure 3.9.

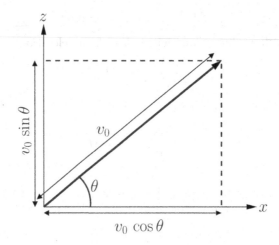

Figure 3.9 Coordinates for the projectile problem

Because the projectile moves with constant acceleration, its vector displacement $\mathbf{s} = (x, y, z)$ from its launch point satisfies [see Equation (3.62)]

$$\mathbf{s} = \mathbf{v}_0\, t + \frac{1}{2}\,\mathbf{a}\, t^2. \tag{3.67}$$

Making use of Equations (3.65) and (3.66), the x-, y-, and z-components of the previous equation are written

$$x = v_0\, \cos\theta\ t, \tag{3.68}$$

$$y = 0, \tag{3.69}$$

$$z = v_0\, \sin\theta\ t - \frac{1}{2}\,g\,t^2, \tag{3.70}$$

respectively. Note that the projectile moves with constant velocity, $v_x \equiv dx/dt = v_0\, \cos\theta$, in the x-direction (i.e., horizontally). This is hardly surprising, because there is zero component of the projectile's acceleration along the x-axis. Note, further, that because there is zero component of the projectile's acceleration along the y-axis, and the projectile's initial velocity also has zero component along this axis, the projectile never moves in the y-direction. In other words, the projectile's trajectory is two-dimensional, lying entirely within the x-z plane. Note, finally, that the projectile's vertical motion is entirely decoupled from its horizontal motion. In other words, the projectile's vertical motion is identical to that of a second projectile launched vertically upward, at $t = 0$, with the initial velocity $v_0\, \sin\theta$ (i.e., the initial vertical velocity component of the first projectile); both projectiles will reach the same maximum altitude at the same time, and will subsequently strike the ground simultaneously.

Equations (3.68) and (3.70) can be rearranged to give

$$z = x\, \tan\theta - \frac{1}{2}\,\frac{g\, x^2}{v_0^2}\, \sec^2\theta. \tag{3.71}$$

As was first pointed out by Galileo, and is illustrated in Figure 3.10, this is the equation of a parabola. The horizontal range, R, of the projectile corresponds to its x-coordinate when

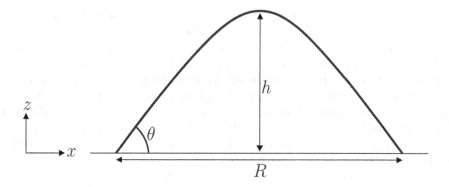

Figure 3.10 The parabolic trajectory of a projectile

it strikes the ground (i.e., when $z = 0$). It follows from the previous expression (neglecting the trivial result $x = 0$) that

$$R = \frac{2\,v_0^2}{g}\,\sin\theta\,\cos\theta = \frac{v_0^2}{g}\,\sin 2\theta. \tag{3.72}$$

Note that the range attains its maximum value,

$$R_{\text{max}} = \frac{v_0^2}{g}, \tag{3.73}$$

when $\theta = 45°$. In other words, neglecting air resistance, a projectile travels furthest when it is launched into the air at 45° to the horizontal.

The maximum altitude, h, of the projectile is attained when $v_z \equiv dz/dt = 0$ (i.e., when the projectile has just stopped rising, and is about to start falling). It follows from Equation (3.70) that the maximum altitude occurs at time $t_0 = v_0 \sin\theta/g$. Hence,

$$h \equiv z(t_0) = \frac{v_0^2}{2\,g}\,\sin^2\theta. \tag{3.74}$$

Obviously, the largest value of h,

$$h_{\text{max}} = \frac{v_0^2}{2\,g}, \tag{3.75}$$

is obtained when the projectile is launched vertically upward (i.e., $\theta = 90°$).

3.7   RELATIVE VELOCITY

Suppose that, on a windy day, an airplane moves with constant velocity, $\mathbf{v}_a$ with respect to the air, and that the air moves with constant velocity, $\mathbf{u}$ with respect to the ground. What is the vector velocity, $\mathbf{v}_g$, of the plane with respect to the ground? In principle, the answer to this question is very simple:

$$\mathbf{v}_g = \mathbf{v}_a + \mathbf{u}. \tag{3.76}$$

In other words, the velocity of the plane with respect to the ground is the vector sum of the plane's velocity relative to the air and the air's velocity relative to the ground. See

Figure 3.11. Note that, in general, $\mathbf{v}_g$ is parallel to neither $\mathbf{v}_a$ nor $\mathbf{u}$. Let us now consider how we might implement Equation (3.76) in practice.

As always, our first task is to set up a suitable Cartesian coordinate system. A convenient system for dealing with two-dimensional motion parallel to the Earth's surface is illustrated in Figure 3.12. The x-axis points northward, whereas the y-axis points eastward. In this coordinate system, it is conventional to specify a vector $\mathbf{r}$ in term of its magnitude, r, and its *compass bearing*, ϕ. A compass bearing is the angle subtended between the direction of a vector and the direction to the north pole; that is, the x-direction. By convention, compass bearings run from $0°$ to $360°$. Furthermore, the compass bearings of north, east, south, and west are $0°$, $90°$, $180°$, and $270°$, respectively.

From simple trigonometry, the components of a general vector $\mathbf{r}$, whose magnitude is r, and whose compass bearing is ϕ, are

$$\mathbf{r} = (x,\, y) = (r\,\cos\phi,\, r\,\sin\phi). \tag{3.77}$$

Note that we have suppressed the z-component of $\mathbf{r}$ (which is zero), for ease of notation.

As an illustration, suppose that the plane's velocity relative to the air is $300\,\text{km/h}$ at a compass bearing of $120°$, and the air's velocity relative to the ground is $85\,\text{km/h}$ at a compass bearing of $225°$. It follows that the components of $\mathbf{v}_a$ and $\mathbf{u}$ (measured in units of km/h) are

$$\mathbf{v}_a = (300\,\cos 120°,\, 300\,\sin 120°) = (-1.500 \times 10^2,\, 2.598 \times 10^2), \tag{3.78}$$

$$\mathbf{u} = (85\,\cos 225°,\, 85\,\sin 225°) = (-6.010 \times 10^1,\, -6.010 \times 10^1). \tag{3.79}$$

According to Equation (3.76), the components of the plane's velocity, $\mathbf{v}_g$, relative to the ground are simply the algebraic sums of the corresponding components of $\mathbf{v}_a$ and $\mathbf{u}$. Hence,

$$\mathbf{v}_g = (-1.500 \times 10^2 - 6.010 \times 10^1,\, 2.598 \times 10^2 - 6.010 \times 10^1)$$
$$= (-2.101 \times 10^2,\, 1.997 \times 10^2). \tag{3.80}$$

Our final task is to reconstruct the magnitude and compass bearing of vector $\mathbf{v}_g$, given its components $(v_{g\,x}, v_{g\,y})$. The magnitude of $\mathbf{v}_g$ is

$$v_g = \sqrt{(v_{g\,x})^2 + (v_{g\,y})^2}$$
$$= \sqrt{(-2.101 \times 10^2)^2 + (1.997 \times 10^2)^2} = 289.9\,\text{km/h}. \tag{3.81}$$

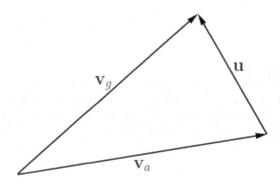

Figure 3.11 Relative velocity

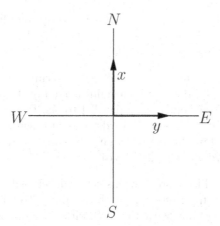

Figure 3.12 Coordinates for the relative velocity problem

In principle, the compass bearing of $\mathbf{v}_g$ is given by the following formula:

$$\phi = \tan^{-1}\left(\frac{v_{g\,y}}{v_{g\,x}}\right). \tag{3.82}$$

This follows because $v_{g\,x} = v_g\cos\phi$ and $v_{g\,y} = v_g\sin\phi$. [See Equation (3.77).] Unfortunately, the previous expression is ambiguous, because it possesses two solutions. An unambiguous pair of expressions for ϕ is as follows:

$$\phi = \tan^{-1}\left(\frac{v_{g\,y}}{v_{g\,x}}\right) \tag{3.83}$$

when $v_{g\,x} \geq 0$; or

$$\phi = 180° - \tan^{-1}\left(\frac{v_{g\,y}}{|v_{g\,x}|}\right) \tag{3.84}$$

when $v_{g\,x} < 0$. These expressions can be derived from simple trigonometry. In both expressions, $-90° \leq \tan^{-1}\theta \leq 90°$. For the case in hand, Equation (3.84) is the relevant expression. Hence,

$$\phi = 180° - \tan^{-1}\left(\frac{1.997 \times 10^2}{2.101 \times 10^2}\right) = 136.5°. \tag{3.85}$$

Thus, the plane's velocity relative to the ground is 289.9 km/h at a compass bearing of 136.5°.

3.8 EXERCISES

3.1 If $\mathbf{c} = \mathbf{a} + \mathbf{b}$, where $\mathbf{a}$, $\mathbf{b}$, and $\mathbf{c}$ are general vectors, demonstrate that $|\mathbf{c}| \leq |\mathbf{a}| + |\mathbf{b}|$.

3.2 An object moves with constant velocity on the straight-line

$$\mathbf{r} = \mathbf{r}_0 + \mathbf{v}\,t.$$

Here, $\mathbf{r}_0$ and $\mathbf{v}$ are constant vectors. Consider a second stationary object located at position vector $\mathbf{r}_1$. Show that the two objects are closest together at time

$t = -[(\mathbf{r}_0 - \mathbf{r}_1) \cdot \hat{\mathbf{v}}]/v$, where $\hat{\mathbf{v}} = \mathbf{v}/v$ is a unit-length vector directed parallel to the first object's velocity vector. Show that the distance of closest approach is $\sqrt{|\mathbf{r}_0 - \mathbf{r}_1|^2 - [(\mathbf{r}_0 - \mathbf{r}_1) \cdot \hat{\mathbf{v}}]^2}$.

3.3 A quarterback receives the snap at the line of scrimmage, takes a seven step drop (i.e., runs backward 9 yards), but is then flushed out of the pocket by a blitzing safety. The quarterback subsequently runs parallel to the line of scrimmage for 12 yards, and then gets off a forward pass, 36 yards straight downfield, to a wide receiver, just prior to being tackled by the safety. What is the magnitude of the football's resultant displacement (in yards)? [Ans: 29.55 yrd.]

3.4 Legend has it that Galileo tested out his newly developed theory of projectile motion by throwing weights from the top of the Leaning Tower of Pisa. (No wonder he eventually got into trouble with the authorities!) Suppose that, one day, Galileo simultaneously threw two equal weights off the tower from a height of 100 m above the ground. Suppose, further, that he dropped the first weight straight down, whereas he threw the second weight horizontally with a velocity of 5 m/s. Neglect the effect of air resistance.

   (a) Which weight struck the ground first? [Ans: Both weights struck the ground at the same time.]

   (b) How long, after it was thrown, did it take to do this? [Ans: 4.515 s.]

   (c) What horizontal distance was traveled by the second weight before it hit the ground? [Ans: 22.58 m.]

3.5 A cannon placed on a 50 m high cliff fires a cannonball over the edge of the cliff at $v = 200$ m/s making an angle of $\theta = 30°$ to the horizontal. How long is the cannonball in the air? Neglect air resistance. [Ans: 20.88 s.]

3.6 A fort is on the edge of a cliff of height h. Show that the greatest horizontal distance at which a gun on the cliff can hit a ship is $2\sqrt{k(k + h)}$, whereas the greatest horizontal distance at which a gun on the ship can hit the fort is $2\sqrt{k(k - h)}$. Here, $\sqrt{2gk}$ is the muzzle-velocity of the shot in each case. Neglect air resistance. [From Lamb 1942.]

3.7 A gun is located at the bottom of a hill of constant slope α, measured with respect to the horizontal. Let θ be the angle of elevation of the gun with respect to the horizontal. Let v be the initial speed of the gun's projectile. Neglect air resistance.

   (a) Show that the range of the gun, measured up the hill, is

$$\frac{2v^2 \cos\theta \, \sin(\theta - \alpha)}{g \cos^2\alpha}.$$

   (b) Show that the maximum value of the slope range is

$$\frac{v^2}{g(1 + \sin\alpha)}.$$

[From Fowles & Cassiday 2005.]

3.8 A projectile is launched from the ground with velocity v_0 at an angle θ to the horizontal. The projectile passes through two points that are both a height h above the ground. Neglect air resistance. Show that the horizontal separation of the two points is

$$\frac{2\,v_0}{g}\,\cos\theta\sqrt{v_0^2\,\sin^2\theta - 2\,g\,h}.$$

3.9 A particle is projected at an elevation θ with respect to an incline of elevation α. The particle subsequently strikes the incline at right angles. Prove that $\tan\theta = (1/2)\cot\alpha$. [From Lamb 1942.]

3.10 Particles are projected simultaneously from a point, in different directions, with equal speeds v. Show that after t seconds the particles will lie on the surface of a sphere of radius $v\,t$ whose center is accelerating downward with acceleration g. [From Lamb 1942.]

3.11 Your favorite football team is down by 4 points with 5 s left in the fourth quarter. The quarterback launches a Hail Mary pass into the end-zone, 60 yards away, where a wide receiver is waiting to make the catch. Suppose that the quarterback throws the ball at 55 miles per hour. At what angle to the horizontal must the ball be launched in order for it to hit the receiver? Neglect the effect of air resistance. [Ans: Either 31.45° or 58.56°.]

3.12 An airplane is 20 miles due north of an airport. Suppose that the plane is flying at 200 mi/h relative to the air. Suppose, further, that there is a wind blowing due east at 60 mi/h. Toward which compass bearing must the plane steer in order to land at the airport? [Ans: 197.46°.]

3.13 One ship is approaching a port at speed u, and another is leaving at speed v, the courses being straight and making an angle α with one another. Show that when the distance between the two ships is least their distances from port are in the ratio $v + u\cos\alpha$ to $u + v\cos\alpha$, respectively. [From Lamb 1942.]

CHAPTER 4

Newton's Laws of Motion

4.1 INTRODUCTION

In his *Principia*, Newton reduced the basic principles of mechanics to three laws:

1. Every body continues in its state of rest, or uniform motion in a straight-line, unless compelled to change that state by forces impressed upon it.

2. The change of motion of an object is proportional to the force impressed upon it, and is made in the direction of the straight-line in which the force is impressed.

3. To every action there is always opposed an equal reaction; or, the mutual actions of two bodies upon each other are always equal and directed to contrary parts.

These laws are known as **Newton's first law of motion**, **Newton's second law of motion**, and **Newton's third law of motion**, respectively. In this chapter, we shall examine each of these laws in detail, and then give some illustrations of their use.

4.2 NEWTON'S FIRST LAW OF MOTION

Newton's first law was actually discovered by Galileo, and perfected by Descartes (who added the crucial proviso "in a straight-line"). This law states that if the motion of a given body is not disturbed by any external influences then that body moves with constant velocity. In other words, the displacement, $\mathbf{r}$, of the body as a function of time, t, can be written

$$\mathbf{r} = \mathbf{r}_0 + \mathbf{v}\,t, \qquad (4.1)$$

where $\mathbf{r}_0$ and $\mathbf{v}$ are constant vectors. As illustrated in Figure 3.8, the body's trajectory is a straight-line that passes through point $\mathbf{r}_0$ at time $t = 0$, and runs parallel to $\mathbf{v}$. In the special case in which $\mathbf{v} = \mathbf{0}$, the body simply remains at rest.

Nowadays, Newton's first law strikes us as almost a statement of the obvious. However, in Galileo's time, this was far from being the case. From the time of the ancient Greeks, philosophers—observing that objects set into motion on the Earth's surface eventually come to rest—had concluded that the natural state of motion of objects was that they should remain at rest. Hence, they reasoned, any object that moves does so under the influence of an external influence, or force, exerted upon it by some other object in the universe. It took the genius of Galileo to realize that an object set into motion on the Earth's surface eventually comes to rest under the influence of frictional forces, and that, if these forces could somehow be abstracted from the motion then the motion would continue forever.

DOI: 10.1201/9781003198642-4

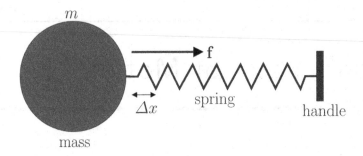

Figure 4.1 Hooke's law

4.3 NEWTON'S SECOND LAW OF MOTION

Newton used the word "motion" to mean what we nowadays call *momentum*. The momentum, **p**, of a body is defined as the product of its mass, m, and its velocity, **v**; that is,

$$\mathbf{p} = m\,\mathbf{v}. \tag{4.2}$$

Newton's second law of motion is summed up in the equation

$$\frac{d\mathbf{p}}{dt} = \mathbf{f}, \tag{4.3}$$

where the vector **f** represents the net influence, or **force**, exerted on the object, whose motion is under investigation by other objects in the universe. For the case of a object with constant mass, the previous law reduces to its more conventional form

$$m\,\mathbf{a} = \mathbf{f}. \tag{4.4}$$

In other words, the product of a given object's mass and its acceleration, $\mathbf{a} = d\mathbf{v}/dt$, is equal to the net force exerted on that object by the other objects in the universe. Of course, this law is entirely devoid of content unless we have some independent means of quantifying the forces exerted between different objects.

4.4 MEASUREMENT OF FORCE

One method of quantifying the force exerted on an object is via *Hooke's law*. (See Section 5.6.) This law—discovered by the English scientist Robert Hooke in 1660—states that the force, f, exerted by a coiled spring is directly proportional to its extension, Δx. The extension of the spring is the difference between its actual length and its natural length (i.e., its length when it is exerting no force).

The force acts parallel to the axis of the spring. In fact, Hooke's law only holds if the extension of the spring is sufficiently small. If the extension becomes too large then the spring deforms permanently, or even breaks. Such behavior lies beyond the scope of Hooke's law.

Figure 4.1 illustrates how we might use Hooke's law to quantify the force that we exert on a body of mass m when we pull on the handle of a spring attached to it. The magnitude, f, of the force is proportional to the extension of the spring; twice the extension means twice the force. As shown, the direction of the force is toward the spring, parallel to its axis (assuming that the extension is positive). The magnitude of the force can be quantified

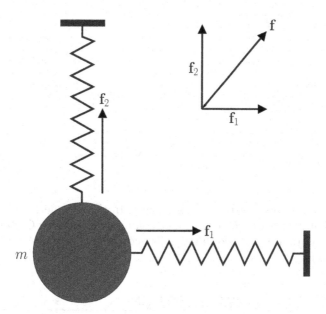

Figure 4.2 Addition of forces

in terms of the critical extension required to impart a unit acceleration (i.e., $1\,\text{m/s}^2$) to a body of unit mass (i.e., $1\,\text{kg}$). According to Equation (4.4), the force corresponding to this extension is $1\,\text{N}$. Thus, if the critical extension corresponds to a force of $1\,\text{N}$ then half the critical extension corresponds to a force of $0.5\,\text{N}$, and so on. In this manner, we can quantify both the direction and magnitude of the force that we exert, by means of a spring, on a given body.

Suppose that we apply two forces, $\mathbf{f}_1$ and $\mathbf{f}_2$ (say), acting in different directions, to a body of mass m by means of two springs. As illustrated in Figure 4.2, it is a matter of experience that the body accelerates as if it were subject to a single force, $\mathbf{f}$, that is the vector sum of the individual forces, $\mathbf{f}_1$ and $\mathbf{f}_2$. Assuming that this is a general result, it follows that the force, $\mathbf{f}$, appearing in Newton's second law of motion, Equation (4.4), is the resultant (i.e., the vector sum) of all the external forces to which the body whose motion is under investigation is subject to.

A corollary of the previous result is that if a body is subject to a single force, $\mathbf{f}$, then its acceleration is the same as if it were subject to a number of forces whose vector sum is equal to $\mathbf{f}$. This corollary is the basis for the *resolution of forces*, by which a force pointing in a general direction can be replaced by three mutually perpendicular forces directed along the Cartesian coordinate axes; that is, $\mathbf{f} = f_x\,\mathbf{e}_x + f_y\,\mathbf{e}_y + f_z\,\mathbf{e}_z$, where f_x, f_y, and f_z are the Cartesian components of the force.

Suppose that the resultant of all the forces acting on a given body is zero. In other words, suppose that the forces acting on the body exactly balance one another. According to Newton's second law of motion, Equation (4.4), the body does not accelerate; that is, it either remains at rest, or moves with uniform velocity in a straight-line. It follows that Newton's first law of motion applies not only to bodies that have no forces acting upon them, but also to bodies acted upon by exactly balanced forces.

A corollary of the previous result is that we can use the calibrated spring system pictured in Figure 4.1 to measure the net force acting on a given body, irrespective of the nature of the force. We just need to apply an additional force to the body, by means of the spring, that is such as to reduce the body's acceleration to zero. The net force acting on the

Figure 4.3 Newton's third law

body is then minus the force exerted by the spring. In this manner, it is possible to prove experimentally that all forces are vector quantities [i.e., their three Cartesian components transform under rotation of the coordinate axis according to Equations (3.20)–(3.22), and their generalizations].

4.5 NEWTON'S THIRD LAW OF MOTION

Suppose, for the sake of argument, that there are only two bodies in the universe. Let us label these bodies a and b. Suppose that body b exerts a force, $\mathbf{f}_{ab}$, on body a. According to to Newton's third law of motion, body a must exert an equal and opposite force, $\mathbf{f}_{ba} = -\mathbf{f}_{ab}$, on body b. See Figure 4.3. This is true irrespective of the nature of the force acting between the two bodies. Thus, if we label $\mathbf{f}_{ab}$ the "action" then, in Newton's language, $\mathbf{f}_{ba}$ is the equal and opposed "reaction".

Suppose, now, that there are many objects in the universe (as is, indeed, the case). According to Newton's third law, if object j exerts a force, $\mathbf{f}_{ij}$, on object i then object i must exert an equal and opposite force, $\mathbf{f}_{ji} = -\mathbf{f}_{ij}$, on object j. It follows that all of the forces acting in the universe can ultimately be grouped into equal and opposite action-reaction pairs. Note, incidentally, that an action and its associated reaction always act on different bodies.

Why do we need Newton's third law? Actually, it is almost a matter of common sense. Suppose that bodies a and b constitute an isolated system. If $\mathbf{f}_{ba} \neq -\mathbf{f}_{ab}$ then this system exerts a non-zero net force, $\mathbf{f} = \mathbf{f}_{ab} + \mathbf{f}_{ba}$, on itself, without the aid of any external agency. It will, therefore, accelerate forever under its own steam. We know, from experience, that this sort of behavior does not occur in real life. For instance, a person cannot grab hold of their own shoelaces and, thereby, pick themselves up off the ground. In other words, the person in question cannot self-generate a force that will spontaneously lift them into the air; they need to exert forces on other objects around them in order to achieve this goal. Thus, Newton's third law essentially acts as a guarantee against the absurdity of self-generated forces.

4.6 MASS, WEIGHT, AND REACTION

The terms *mass* and *weight* are often confused with one another. However, in physics, their meanings are quite distinct.

A body's mass is a measure of its *inertia*; that is, its reluctance to deviate from uniform straight-line motion under the influence of external forces. According to Newton's second law, Equation (4.4), if two objects of differing masses are acted upon by forces of the same magnitude then the resulting acceleration of the larger mass is less than that of the smaller

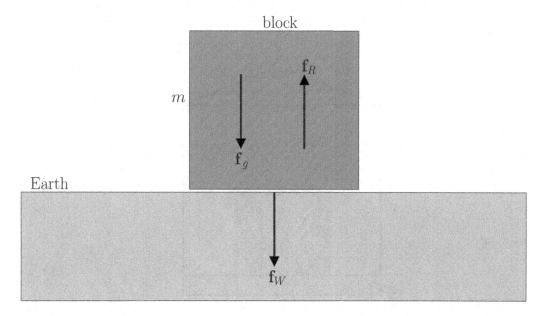

Figure 4.4 Block resting on the surface of the Earth

mass. In other words, it is more difficult to force the larger mass to deviate from its preferred state of uniform motion in a straight-line. Incidentally, the mass of a body is an intrinsic property of that body, and, therefore, does not change if the body is moved to a different location.

4.6.1 Block Resting on Earth's Surface

Imagine a block of granite resting on the surface of the Earth. See Figure 4.4. The block experiences a downward force, $\mathbf{f}_g$, due to the gravitational attraction of the Earth. This force is of magnitude $m\,g$, where m is the mass of the block, and $g = 9.81\,\mathrm{m/s}^{-2}$ is the acceleration due to gravity at the surface of the Earth. The block transmits this force to the ground below it, which is supporting it, and, thereby, preventing it from accelerating downward. In other words, the block exerts a downward force, $\mathbf{f}_W$, of magnitude $m\,g$, on the ground immediately beneath it. We usually refer to this force (or the magnitude of this force) as the weight of the block. According to Newton's third law, the ground below the block exerts an upward *reaction* force, $\mathbf{f}_R$, on the block. This force is also of magnitude $m\,g$. Thus, the net force acting on the block is $\mathbf{f}_g + \mathbf{f}_R = \mathbf{0}$, which accounts for the fact that the block remains stationary.

Where, you might ask, is the equal and opposite reaction to the force of gravitational attraction, $\mathbf{f}_g$, exerted by the Earth on the block of granite? It turns out that this reaction is exerted at the center of the Earth. In other words, the Earth attracts the block of granite, and the block of granite attracts the Earth by an equal amount. However, because the Earth is far more massive than the block, the force exerted by the granite block at the center of the Earth has no observable consequence.

So far, we have established that the weight, W, of a body is the magnitude of the downward force that it exerts on any object which supports it. Thus, $W = m\,g$, where m is the mass of the body, and g is the local acceleration due to gravity. Because weight is a force, it is measured in newtons. A body's weight is location dependent, and is not, therefore, an

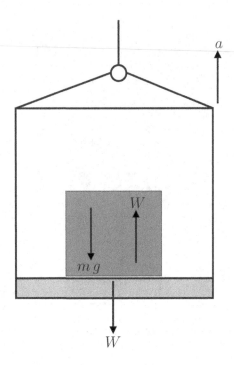

Figure 4.5 Weight in an elevator

intrinsic property of that body. For instance, a body weighing 10 N on the surface of the Earth would only weigh about 3.8 N on the surface of Mars, due to the weaker surface gravity of Mars relative to the Earth. (See Table 13.1.)

4.6.2 Block in an Elevator

Consider a block of mass m resting on the floor of an elevator, as shown in Figure 4.5. Suppose that the elevator is accelerating upward with acceleration a. How does this acceleration affect the weight of the block? Of course, the block experiences a downward force, $m\,g$, due to gravity. Let W be the weight of the block; by definition, this is the magnitude of the downward force exerted by the block on the floor of the elevator. From Newton's third law, the floor of the elevator exerts an upward reaction force of magnitude W on the block. Let us apply Newton's second law, Equation (4.4), to the motion of the block. The mass of the block is m, and its upward acceleration is a. Furthermore, the block is subject to two forces; a downward force, $m\,g$, due to gravity, and an upward reaction force, W. Hence,

$$m\,a = W - m\,g. \tag{4.5}$$

This equation can be rearranged to give

$$W = m\,(g + a). \tag{4.6}$$

Clearly, the upward acceleration of the elevator has the effect of increasing the weight, W, of the block; for instance, if the elevator accelerates upward at g then the weight of the block is doubled. Conversely, if the elevator accelerates downward (i.e., if a becomes negative) then the weight of the block is reduced; for instance, if the elevator accelerates downward at $g/2$

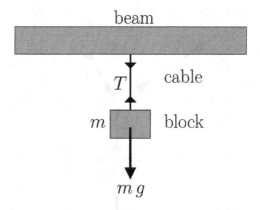

Figure 4.6 Block suspended by a cable

then the weight of the block is halved. Incidentally, these weight changes could easily be measured by placing some bathroom scales between the block and the floor of the elevator.

Suppose that the downward acceleration of the elevator matches the acceleration due to gravity; that is, $a = -g$. In this case, $W = 0$. In other words, the block becomes weightless! This is the principle behind the so-called "Vomit Comet" used by NASA's Johnson Space Center to train prospective astronauts in the effects of weightlessness. The "Vomit Comet" is actually a KC-135 (a predecessor of the Boeing 707 that is usually employed for refueling military aircraft). The plane typically ascends to 30 000 ft, and then accelerates downward at g, in a parabolic path, for about 20 s, allowing its passengers to feel the effects of weightlessness during this time period. All of the weightless scenes in the movie Apollo 11 were shot in this manner.

Suppose, finally, that the downward acceleration of the elevator exceeds the acceleration due to gravity; that is, $a < -g$. In this case, the block acquires a negative weight! What actually happens is that the block flies off the floor of the elevator and slams into the ceiling. When things have settled down, the block exerts an upward force (negative weight) $|W|$ on the ceiling of the elevator.

4.7   SUSPENDED MASSES

4.7.1   Block Suspended by a Single Cable

Consider a block of mass m that is suspended from a fixed beam by means of a cable, as shown in Figure 4.6. The cable is assumed to be light (i.e., its mass is negligible compared to that of the block), and inextensible (i.e., its length increases by a negligible amount because of the weight of the block). The cable is clearly being stretched, because it is being pulled at both ends by the block and the beam. Furthermore, the cable must be being pulled by oppositely directed forces of the same magnitude, otherwise it would accelerate greatly (given that it has negligible inertia). By Newton's third law, the cable exerts oppositely directed forces of equal magnitude, T (say), on both the block and the beam. These forces act so as to oppose the stretching of the cable; that is, the beam experiences a downward force of magnitude T, whereas the block experiences an upward force of magnitude T. Here, T is termed the *tension* of the cable. Because T is a force, it is measured in newtons. Note that, unlike a coiled spring, a cable can never possess a negative tension, because this would imply that the cable is trying to push its supports apart, rather than pull them together.

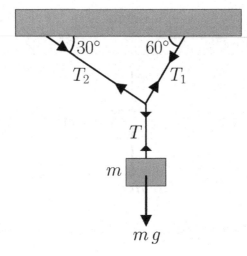

Figure 4.7 Block suspended by three cables

Let us apply Newton's second law to the block. The mass of the block is m, and its acceleration is zero, because the block is assumed to be in equilibrium. The block is subject to two forces, a downward force, $m\,g$, due to gravity, and an upward force, T, due to the tension of the cable. It follows that

$$0 = T - m\,g. \tag{4.7}$$

In other words, in equilibrium, the tension, T, of the cable equals the weight, $m\,g$, of the block.

4.7.2 Block Suspended by Three Cables

Figure 4.7 shows a slightly more complicated example in which a block of mass m is suspended by three cables. What are the tensions, T, T_1, and T_2, in these cables, assuming that the block is in equilibrium? Using analogous arguments to the previous case, we can easily demonstrate that the tension, T, in the lowermost cable is $m\,g$. The tensions in the two uppermost cables are obtained by applying Newton's second law of motion to the junction where all three cables meet.

There are three forces acting on the junction; the downward force, T, due to the tension in the lower cable, and the forces, T_1 and T_2, due to the tensions in the upper cables. The latter two forces act parallel to their respective cables, as indicated in the figure. Because the junction is in equilibrium, the vector sum of all the forces acting on it must be zero.

Consider the horizontal components of the forces acting on the junction. Let components acting to the right be positive, and vice versa. The horizontal component of the tension T is zero, because this tension acts straight down. The horizontal component of the tension T_1 is $T_1 \cos 60° = T_1/2$, because this force subtends an angle of $60°$ with respect to the horizontal. (See Figure 3.9.) Likewise, the horizontal component of the tension T_2 is $-T_2 \cos 30° = -\sqrt{3}\,T_2/2$. Because the junction does not accelerate in the horizontal direction, we can equate the sum of these components to zero:

$$\frac{1}{2}\,T_1 - \frac{\sqrt{3}}{2}\,T_2 = 0. \tag{4.8}$$

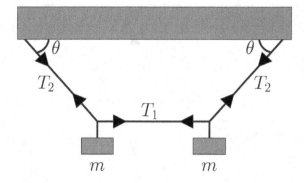

Figure 4.8 Two blocks suspended by five cables

Consider the vertical components of the forces acting on the junction. Let components acting upward be positive, and vice versa. The vertical component of the tension T is $-T = -m\,g$, because this tension acts straight down. The vertical component of the tension T_1 is $T_1\sin 60° = \sqrt{3}\,T_1/2$, because this force subtends an angle of $60°$ with respect to the horizontal. Likewise, the vertical component of the tension T_2 is $T_2\sin 30° = T_2/2$. Because the junction does not accelerate in the vertical direction, we can equate the sum of these components to zero:

$$-m\,g + \frac{\sqrt{3}}{2}\,T_1 + \frac{1}{2}\,T_2 = 0. \tag{4.9}$$

Finally, Equations (4.8) and (4.9) yield the following expressions for the tensions in the two upper cables:

$$T_1 = \frac{\sqrt{3}}{2}\,m\,g, \tag{4.10}$$

$$T_2 = \frac{1}{2}\,m\,g. \tag{4.11}$$

Note that the tension is greater in the more vertical cable. Incidentally, this type of calculation is vital because there is a maximum safe tension that can be applied to a cable of a given diameter, made of a given material, and it is important to ensure that this maximum tension is not exceeded when using such a cable to suspend objects.

4.7.3 Two Blocks Suspended by Five Cables

Consider the situation shown in Figure 4.8 in which two blocks of equal mass, m, are suspended by five cables. Using analogous arguments to the two previous cases, we can easily demonstrate that the common tension, T, in the two vertical cables is $m\,g$.

Consider the two mirror-image junctions in which three cables meet. Because the junctions do not accelerate in the horizontal direction, we can equate the horizontal components of the forces acting at the junctions to give

$$T_2\cos\theta = T_1. \tag{4.12}$$

Likewise, because the junctions do not accelerate in the vertical direction, we can equate the vertical components of the forces acting at the junctions to give

$$T_2\sin\theta = T = m\,g. \tag{4.13}$$

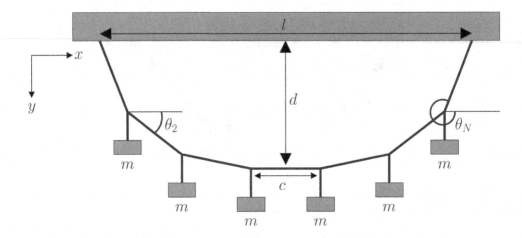

Figure 4.9 Many masses suspended by many cables

Hence, we deduce that

$$T_1 = \frac{m\,g}{\tan\theta}, \tag{4.14}$$

$$T_2 = \frac{m\,g}{\sin\theta} = \left[T_1^2 + (m\,g)^2\right]^{1/2}, \tag{4.15}$$

because $\sin\theta = \tan\theta/(1 + \tan^2\theta)^{1/2}$. As before, the more vertical cables have the higher tensions.

4.7.4   Many Blocks Suspended by Many Cables

Consider the situation illustrated in Figure 4.9 in which a large number of identical blocks of mass, m, are suspended in a chain via identical cables of length, c, such that the two ends of the system are at the same height. Let us, henceforth, refer to the blocks simply as masses, for the sake of generality.

Suppose that there are N masses. Let $i = 1, N$ index the masses. So, the leftmost mass corresponds to $i = 1$, the next mass over corresponds to $i = 2$, and the rightmost mass corresponds to $i = N$. Let $i = 1, N + 1$ index the cables. So, the leftmost cable corresponds to $i = 1$, whereas the rightmost cable corresponds to $i = N + 1$. Let θ_i be the angle that the ith cable subtends with the horizontal, as shown in the figure.

Equating the horizontal forces at the cable junction directly above the ith mass, we find that

$$T_i \cos\theta_i = T_{i+1} \cos\theta_{i+1}. \tag{4.16}$$

Likewise, equating the vertical forces, we find that

$$T_i \sin\theta_i - T_{i+1} \sin\theta_{i+1} = m\,g. \tag{4.17}$$

The previous two equations hold for i in the range 1 to N (because there are N junctions). It follows from Equation (4.16) that

$$T_i = \frac{H}{\cos\theta_i}, \tag{4.18}$$

for $i = 1, N$, where H is the common horizontal component of the tension in each cable. Furthermore, the previous two equations can be combined to give

$$\tan \theta_{i+1} - \tan \theta_i = -\frac{m\,g}{H}, \tag{4.19}$$

for $i = 1, N$. The previous equation specifies a set of N difference equations for $N + 1$ unknowns. However, by symmetry,

$$\tan \theta_{N+1} = -\tan \theta_1. \tag{4.20}$$

Here, θ_1 and θ_{N+1} are the angles subtended between the endmost cables and the supporting beam in Figure 4.9. Hence, there are actually only N unknowns. In fact, it is clear, by inspection, that the solution of the set of difference equations is

$$\tan \theta_i = \frac{m\,g}{H} \left(\frac{N}{2} + 1 - i \right). \tag{4.21}$$

Hence,

$$T_i = \frac{H}{\cos \theta_i} = \left[H^2 + \left(\frac{N}{2} + 1 - i \right)^2 (m\,g)^2 \right]^{1/2}, \tag{4.22}$$

which follows because $\cos \theta = 1/\sqrt{1 + \tan^2 \theta}$. Note that the tension is smallest in the middle (horizontal) cable span, and rises toward the two ends of the system.

Let x and y be horizontal and vertical coordinates, as indicated in Figure 4.9, and let the origin correspond to the suspension point of the leftmost cable. It follows that the x and y coordinates of the ith cable junction are

$$x_i = c \sum_{j=1,i} \cos \theta_j, \tag{4.23}$$

$$y_i = c \sum_{j=1,i} \sin \theta_j, \tag{4.24}$$

respectively. Furthermore, $i = 0$ and $i = N + 1$ correspond to the suspension points of the leftmost and rightmost cable, respectively. Unfortunately, the previous two expressions are difficult to sum exactly.

4.7.5 Catenary

Let us investigate the problem considered in the previous subsection further. We can define $s = i\,c$, for $i = 0, N+1$. This quantity represents length measured along the cables segments, starting from the suspension point of the leftmost cable. In the limit in which the number of masses tends to infinity, and the length, c, of each cable segment tends to zero, our problem transforms into the problem of the suspension of a uniform cable of weight per unit length $w = m\,g/c$ that is suspended from two points of equal height. Furthermore, we can think of s as a continuous variable that measures arc-length along the cable. s runs from 0 at the left end of the cable to L at the right end, where L is the total length of the cable.

Equations (4.23) and (4.24) morph into

$$x(s) = \int_0^s \cos \theta(s')\, ds' \tag{4.25}$$

$$y(s) = \int_0^s \sin \theta(s')\, ds', \tag{4.26}$$

respectively, where x and y are the Cartesian coordinates of the cable, and $\theta(s)$ is the angle that the tangent to the cable at arc-length s subtends with the horizontal.

Equations (4.18) and (4.19) morph into

$$T = \frac{H}{\cos\theta}, \tag{4.27}$$

$$\frac{d\tan\theta}{ds} = -\frac{mg}{Hc} = -\frac{w}{H} = -\frac{1}{a}, \tag{4.28}$$

where

$$a = \frac{H}{w} \tag{4.29}$$

is a constant with the dimensions of length. The boundary condition is [cf., Equation (4.20)]

$$\tan\theta(s = L) = -\tan\theta(s = 0). \tag{4.30}$$

The solution to Equation (4.28), subject to the previous boundary condition, is

$$\tan\theta = \frac{L - 2s}{2a}, \tag{4.31}$$

which is the continuum version of Equation (4.21).

Let $t = \tan\theta$. Equation (4.28) yields $dt/ds = -1/a$. Hence, Equation (4.25) can be transformed to give

$$x(\tan\theta) = \int_{L/(2a)}^{\tan\theta} \frac{1}{\sqrt{1 + t'^2}} \frac{ds}{dt'} dt' = -a \int_{L/(2a)}^{\tan\theta} \frac{dt'}{\sqrt{1 + t'^2}}, \tag{4.32}$$

because $\cos\theta = 1/\sqrt{1 + \tan^2\theta}$. However,

$$\int \frac{dt}{\sqrt{1 + t^2}} = \sinh^{-1} t. \tag{4.33}$$

Thus, we obtain

$$x(\tan\theta) = a \left[\sinh^{-1}\left(\frac{L}{2a}\right) - \sinh^{-1}(\tan\theta) \right]. \tag{4.34}$$

Now,

$$l \equiv x \left(\tan\theta = -\frac{L}{2a} \right) \tag{4.35}$$

is the horizontal distance between the two end points of the cable, as indicated in Figure 4.9, which implies that

$$l = 2a \sinh^{-1}\left(\frac{L}{2a}\right), \tag{4.36}$$

or

$$\frac{L}{2a} = \sinh\left(\frac{l}{2a}\right). \tag{4.37}$$

Equations (4.34) and (4.37) can be combined to give

$$\tan\theta = \sinh\left(\frac{l - 2x}{2a}\right). \tag{4.38}$$

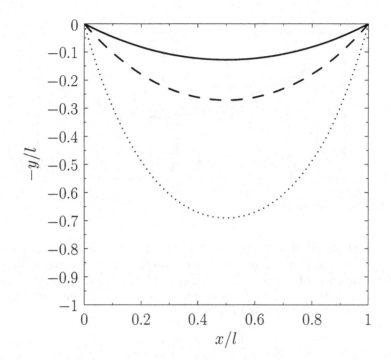

Figure 4.10 Catenary curves. The solid, dashed, and dotted curves correspond to $l/a = 1$, 2, and 4, respectively

Finally, Equation (4.27) produces

$$T = \frac{H}{\cos \theta} = H\sqrt{1 + \tan^2 \theta}, \tag{4.39}$$

which yields

$$T = H \cosh\left(\frac{l - 2x}{2a}\right), \tag{4.40}$$

with the aid of Equation (4.38), and the identity $\cosh x = \sqrt{1 + \sinh^2 x}$.

Equation (4.26) transforms to give

$$y(\tan \theta) = -a \int_{L/(2a)}^{\tan \theta} \frac{t' \, dt'}{\sqrt{1 + t'^2}} = -a \left[\sqrt{1 + t^2}\right]_{L/(2a)}^{\tan \theta}, \tag{4.41}$$

because $\sin \theta = \tan \theta/\sqrt{1 + \tan^2 \theta}$. It follows that

$$y(x) = a \left[\cosh\left(\frac{l}{2a}\right) - \cosh\left(\frac{l - 2x}{2a}\right)\right], \tag{4.42}$$

where use has been made of Equations (4.37) and (4.38), as well as the identity $\cosh x = \sqrt{1 + \sinh^2 x}$. Let

$$d \equiv y(x = l/2) \tag{4.43}$$

be the maximum sag in the cable, as shown in Figure 4.9. It follows from Equation (4.42) that

$$d = a \left[\cosh\left(\frac{l}{2a}\right) - 1\right]. \tag{4.44}$$

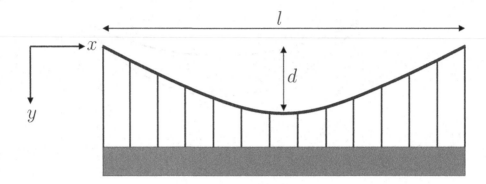

Figure 4.11 A suspension bridge

The curve $y(x)$, specified in Equation (4.42), is known as a *catenary*, after the latin for "chain" (catēna). See Figure 4.10. This curve specifies how a uniform cable (or chain), suspended from supports of the same height, sags under its own weight. Given [see Equation (4.29)] that $a = H/w$, where H is the horizontal tension in the cable, and w is the cable's weight per unit length, the sag can be written

$$d = \frac{H}{w} \left[\cosh\left(\frac{l\,w}{2\,H}\right) - 1 \right]. \tag{4.45}$$

In the limit that the sag is relatively small (i.e., $l \ll a$), the previous equation reduces to

$$d \simeq \frac{l^2\,w}{8\,H} \left[1 + \frac{1}{48}\left(\frac{l\,w}{H}\right)^2 + \cdots \right] \tag{4.46}$$

(because $\cosh x \simeq 1 + x^2/2 + x^4/24$ when $|x| \ll 1$). Hence, it is clear that increasing the horizontal tension reduces the sag. However, the horizontal tension would need to be infinite to completely eliminate the sag.

If the cables in Figure 4.9 are replaced by struts that can withstand a compressive force (i.e., a negative tension) then the system can be inverted to produce an arch. We deduce, from the analysis of this subsection, that an arch of uniform mass per unit length that possesses a uniform horizontal compression takes the form of an inverted catenary. The St Louis (Missouri) Gateway Arch is an example of such an arch.

4.7.6 Suspension Bridge

Figure 4.11 shows a suspension bridge in which a uniform horizontal deck is suspended from a cable. Suppose that the mass of the cable is negligible compared to that of the deck. Let x measure horizontal distance from the suspension point of the left end of the cable, and let y measure vertical distance below the heights of the suspension points of the left and right ends of the cable (which are equal). Let l be the horizontal distance between the suspension points. Finally, let d be the maximum sag in the cable.

The situation under investigation in this subsection is very similar to that investigated in Section 4.7.4, except that now the cable is subject to a uniform load per unit horizontal length, rather than a uniform load per unit arc-length. We can modify the analysis of Section 4.7.4 to take this into account by saying that the length of the ith cable is

$$c_i = \frac{b}{\cos\theta_i}, \tag{4.47}$$

where

$$w = \frac{m\,g}{b} \tag{4.48}$$

is the uniform load per unit horizontal length to which the cable is subject. The analysis of Section 4.7.4 is unchanged except that Equations (4.23) and (4.24) become

$$x_i = i\,b, \tag{4.49}$$

$$y_i = b \sum_{j=1,i} \tan\theta_j. \tag{4.50}$$

If we now repeat the analysis of Section 4.7.5, by taking the continuum limit, we find that

$$y(x) = \int_0^x \tan\theta(x')\,dx'. \tag{4.51}$$

Moreover, Equations (4.27) and (4.28) become

$$T = \frac{H}{\cos\theta}, \tag{4.52}$$

$$\frac{d\tan\theta}{dx} = -\frac{m\,g}{H\,b} = -\frac{w}{H} = -\frac{1}{a}, \tag{4.53}$$

where H is the horizontal tension in the cable, and

$$a = \frac{H}{w} \tag{4.54}$$

is a constant with the dimensions of length. The boundary condition is

$$\tan\theta(x = l) = -\tan\theta(x = 0). \tag{4.55}$$

The solution to Equation (4.53), subject to the previous boundary condition, is

$$\tan\theta = \frac{l - 2\,x}{2\,a}. \tag{4.56}$$

Hence, Equation (4.51) yields

$$y(x) = \int_0^x \left(\frac{l - 2\,x'}{2\,a}\right) dx' = \frac{x\,(l - x)}{2\,a}. \tag{4.57}$$

It is evident that the cable subject to a uniform load per horizontal length adopts a parabolic shape.

The maximum sag in the cable is $d = y(l/2)$, or

$$d = \frac{l^2\,w}{8\,H}. \tag{4.58}$$

It follows from Equations (4.54) and (4.57) that

$$y(x) = 4\,d \left(\frac{x}{l}\right) \left(1 - \frac{x}{l}\right). \tag{4.59}$$

Moreover, Equation (4.52) implies that the tension in the cable is

$$T(x) = \frac{w\,l}{2} \left[\left(1 - \frac{2\,x}{d}\right)^2 + \frac{l^2}{16\,d^2}\right]^{1/2}, \tag{4.60}$$

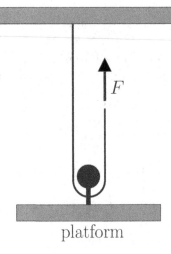

platform

Figure 4.12 A simple pulley

where use has been made of Equation (4.56), as well as the identity $1/\cos\theta = \sqrt{1+\tan^2\theta}$. As usual, the tension is lowest in the middle of the cable, and attains a maximum value

$$T_{\max} = \frac{w\,l}{2}\left(1+\frac{l^2}{16\,d^2}\right)^{1/2} \tag{4.61}$$

at the two ends of the cable.

The arc-length of the cable is

$$L = \int_0^l \left[1+\left(\frac{dy}{dx}\right)^2\right]^{1/2} dx = \int_0^l \left[1+\left(\frac{4\,d}{l}\right)^2\left(1-\frac{2\,x}{l}\right)^2\right]^{1/2} dx, \tag{4.62}$$

where use has been made of Equation (4.57). Expanding in d/l, the previous expression yields

$$L = l\left[1+\frac{8}{3}\left(\frac{d}{l}\right)^2 - \frac{32}{5}\left(\frac{d}{l}\right)^4 + \frac{256}{7}\left(\frac{d}{l}\right)^6 + \cdots\right]. \tag{4.63}$$

4.8 CABLE-PULLEY SYSTEMS

4.8.1 Simple Pulley

Consider the situation pictured in Figure 4.12. A movable platform of mass m has a light frictionless pulley attached to it. A cable is run around the pulley. One end of the cable is attached to a fixed support above the platform, whereas the other end its pulled upward with a force F. Let us calculate the vertical acceleration of the platform.

Because the pulley is light and frictionless, the tensions in the two vertical legs of the cable must be equal, otherwise the pulley would be subject to an net torque that would cause an infinite angular acceleration. Let T denote the common tensions. By Newton's third law of motion, $T = F$.

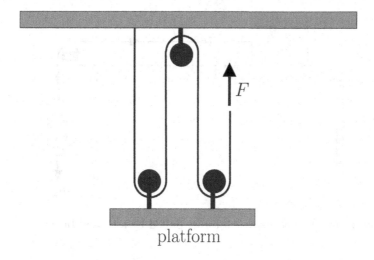

Figure 4.13 A compound pulley

Because the pulley is light, no net force can be acting on it (otherwise it would have an infinite linear acceleration). Thus, the force that the pulley exerts on the platform must equal the net force exerted on it by the two vertical cable legs, which is $2\,T$.

Consider the vertical equation of motion of the platform. Let a be the platform's upward acceleration. It follows that

$$m\,a = 2\,T - m\,g. \tag{4.64}$$

Suppose that the platform is moving upward at a constant velocity, so that $a = 0$. It follows that $T = m\,g/2$, and, hence, that $F = m\,g/2$. In other words, the upward force that must be exerted on the free end of the cable, in order to slowly raise the platform, is only half the weight of the platform. Hence, it is advantageous to lift the platform by means of the simple pulley system shown in the figure, rather than directly, because the necessary force that must be exerted is reduced by a factor of two.

4.8.2 Compound Pulley

Consider the situation pictured in Figure 4.13. A movable platform of mass m has two light frictionless pulleys attached to it. A third pulley is attached to a fixed support above the platform. A cable is run around the three pulleys, as shown. One end of the cable is attached to the fixed support, whereas the other end its pulled upward with a force F.

Using analogous arguments to those employed in the previous subsection, the vertical equation of motion of the platform is

$$m\,a = 4\,T - m\,g. \tag{4.65}$$

Suppose that the platform is moving upward at a constant velocity, so that $a = 0$. It follows that $T = m\,g/4$, and, hence, that $F = m\,g/4$. It follows that the upward force that must be exerted on the free end of the cable, in order to slowly raise the platform, is now only one quarter of the weight of the platform. Clearly, the compound pulley system pictured in Figure 4.13 possesses a greater *mechanical advantage* (i.e., force amplification factor) than the simple pulley system pictured in Figure 4.12. Obviously, it is possible to devise a system consisting of even more pulleys that possesses an even greater mechanical advantage.

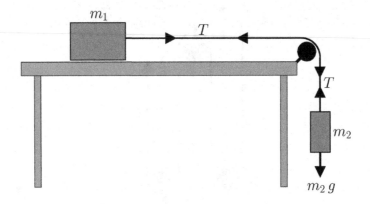

Figure 4.14 Block sliding over a smooth table, pulled by a second block

4.8.3 Table Pulley

Consider two masses, m_1 and m_2, connected by a light inextensible cable. Suppose that the first mass slides over a smooth, frictionless, horizontal table, while the second is suspended over the edge of the table by means of a light frictionless pulley. See Figure 4.14. Because the pulley is light, we can neglect its rotational inertia in our analysis. Moreover, no force is required to turn a frictionless pulley, so we can assume that the tension, T, of the cable is the same on either side of the pulley. Let us apply Newton's second law of motion to each mass in turn. The first mass is subject to a downward force, $m_1 g$, due to gravity. However, this force is completely canceled out by the upward reaction force due to the table. The mass m_1 is also subject to a horizontal force, T, due to the tension in the string, which causes it to move rightward with acceleration

$$a = \frac{T}{m_1}. \tag{4.66}$$

The second mass is subject to a downward force, $m_2 g$, due to gravity, plus an upward force, T, due to the tension in the cable. These forces cause the mass to move downward with acceleration

$$a = g - \frac{T}{m_2}. \tag{4.67}$$

Now, the rightward acceleration of the first mass must match the downward acceleration of the second, because the cable that connects them is inextensible. Thus, equating the previous two expressions, we obtain

$$T = \frac{m_1 m_2}{m_1 + m_2} g, \tag{4.68}$$

$$a = \frac{m_2}{m_1 + m_2} g. \tag{4.69}$$

Note that the acceleration of the two coupled masses is less than the full acceleration due to gravity, g, because the first mass contributes to the inertia of the system, but does not contribute to the downward gravitational force that sets the system into motion.

4.8.4 Atwood Machine

Consider two masses, m_1 and m_2, connected by a light inextensible cable that is suspended from a light frictionless pulley, as shown in Figure 4.15. Let us again apply Newton's second law to each mass in turn. Without being given the values of m_1 and m_2, we cannot determine beforehand which mass is going to move upward. Let us assume that mass m_1 is going to move upward; if we are wrong in this assumption then we will simply obtain a negative acceleration for this mass. The first mass is subject to an upward force, T, due to the tension in the cable, and a downward force, $m_1\,g$, due to gravity. These forces cause the mass to move upward with acceleration

$$a = \frac{T}{m_1} - g. \tag{4.70}$$

The second mass is subject to a downward force, $m_2\,g$, due to gravity, and an upward force, T, due to the tension in the cable. These forces cause the mass to move downward with acceleration

$$a = g - \frac{T}{m_2}. \tag{4.71}$$

Now, the upward acceleration of the first mass must match the downward acceleration of the second, because they are connected by an inextensible cable. Hence, equating the previous two expressions, we obtain

$$T = \frac{2\,m_1\,m_2}{m_1 + m_2}\,g, \tag{4.72}$$

$$a = \frac{m_2 - m_1}{m_1 + m_2}\,g. \tag{4.73}$$

As expected, the first mass accelerates upward (i.e., $a > 0$) if $m_2 > m_1$, and vice versa. Note that the acceleration of the system is less than the full acceleration due to gravity, g, because both masses contribute to the inertia of the system, but their weights partially cancel one another out. In particular, if the two masses are almost equal then the acceleration of the system becomes very much less than g.

Incidentally, the device pictured in Figure 4.15 is called an *Atwood machine*, after the 18th-century English scientist George Atwood, who used it to "slow down" free-fall sufficiently to make accurate observations of this phenomena using the primitive time-keeping devices available in his day.

4.9 VELOCITY-DEPENDENT FORCES

Consider a point particle of mass m moving in one dimension under the action of a force, f, that is a function of the particle's speed, v, but not of its displacement, x. According to Newton's second law of motion, the particle's equation of motion can be written

$$m\,\frac{dv}{dt} = f(v). \tag{4.74}$$

Here, we have expressed the particle's acceleration as the time derivative of its velocity. Integrating this equation, we obtain

$$\int_{v_0}^{v} \frac{dv'}{f(v')} = \frac{t}{m}, \tag{4.75}$$

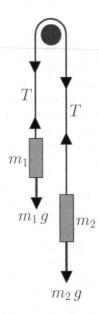

Figure 4.15 An Atwood machine

where $v(t = 0) = v_0$. In principle, the previous equation can be solved to give $v(t)$. The particle's equation of motion can also be written

$$m\,v\,\frac{dv}{dx} = f(v),\tag{4.76}$$

because $v = dx/dt$. Integrating this equation, we obtain

$$\int_{v_0}^{v} \frac{v'\,dv'}{f(v')} = \frac{x - x_0}{m},\tag{4.77}$$

where $x(t = 0) = x_0$. In principle, the previous equation can also be solved to give $v(x)$.

Let us now consider a specific example. Suppose that an object of mass m falls vertically under gravity. Let x be the height through which the object has fallen since $t = 0$, at which time the object is assumed to be at rest. It follows that $x_0 = v_0 = 0$. Suppose that, in addition to the force of gravity, $m\,g$, where g is the gravitational acceleration, our object is subject to a retarding air resistance force that is proportional to the square of its instantaneous velocity. The object's equation of motion is thus

$$m\,\frac{dv}{dt} = f(v),\tag{4.78}$$

where

$$f(v) = m\,g - c\,v^2,\tag{4.79}$$

and $c > 0$. According to Equation (4.75), we can write

$$\int_{0}^{v} \frac{dv'}{m\,g - c\,v'^2} = \frac{t}{m},\tag{4.80}$$

which yields

$$\int_{0}^{v} \frac{dv'}{1 - (v'/v_t)^2} = g\,t,\tag{4.81}$$

where $v_t = (m\,g/c)^{1/2}$. Making a change of variable, we obtain

$$\int_0^{v/v_t} \frac{dy}{1 - y^2} = \frac{g}{v_t}\,t. \tag{4.82}$$

The left-hand side of the previous equation is now a standard integral, which can be performed to give

$$\tanh^{-1}\left(\frac{v}{v_t}\right) = \frac{g\,t}{v_t}, \tag{4.83}$$

or

$$v = v_t \tanh\left(\frac{g\,t}{v_t}\right). \tag{4.84}$$

Thus, when $t \ll v_t/g$, we obtain the standard result $v \simeq g\,t$, because $\tanh x \simeq x$ for $x \ll 1$. However, when $t \gg v_t/g$, we get $v \simeq v_t$, because $\tanh x \simeq 1$ for $x \gg 1$. It follows that air resistance prevents the downward velocity of our object from increasing indefinitely as it falls. Instead, at large times, the velocity asymptotically approaches the so-called *terminal velocity*, v_t (at which the gravitational and air resistance forces balance).

The equation of motion of our falling object is also written

$$m\,v\,\frac{dv}{dx} = f(v), \tag{4.85}$$

where $f(v)$ is specified in Equation (4.79). According to Equation (4.77),

$$\int_0^v \frac{v'\,dv'}{m\,g - c\,v'^2} = \frac{x}{m}, \tag{4.86}$$

which reduces to

$$\int_0^v \frac{v'\,dv'}{1 - (v'/v_t)^2} = g\,x. \tag{4.87}$$

Making a change of variable, we obtain

$$\int_0^{(v/v_t)^2} \frac{dy}{1 - y} = \frac{x}{x_t}, \tag{4.88}$$

where $x_t = m/(2\,c)$. The left-hand side of the previous equation is now a standard integral, which can be performed to give

$$-\ln\left[1 - \left(\frac{v}{v_t}\right)^2\right] = \frac{x}{x_t}, \tag{4.89}$$

or

$$v = v_t \left(1 - e^{-x/x_t}\right)^{1/2}. \tag{4.90}$$

It follows that our object needs to fall a distance of order x_t before it achieves its terminal velocity.

Incidentally, it is quite easy to account for an air resistance force that scales as the square of projectile velocity. Let us imaging that our projectile is moving sufficiently rapidly that air does not have enough time to flow around it, and is instead simply knocked out of the way. If our projectile has cross-sectional area A, perpendicular to the direction of its motion, and is moving with speed v, then the mass of air that it knocks out of its way per second is $\rho_a \, A \, v$, where ρ_a is the mass density of air. Suppose that the air knocked out of the way is pushed in the direction of the projectile's motion with a speed of order v. It follows that the air gains momentum per unit time $\rho_a \, A \, v^2$ in the direction of the projectile's motion. Hence, by Newton's third law of motion, the projectile loses the same momentum per unit time in the direction of its motion. In other words, the projectile is subject to a drag force of magnitude

$$f_{\text{drag}} = C \, \rho_a \, A \, v^2 \tag{4.91}$$

acting in the opposite direction to its motion. Here, C is an $\mathcal{O}(1)$ dimensionless constant, known as the *drag coefficient*, that depends on the exact shape of the projectile. Obviously, streamlined projectiles, such as arrows, have small drag coefficients, whereas non-streamlined projectiles, such as bricks, have large drag coefficients. From before, the terminal velocity of our projectile is $v_t = (m \, g/c)^{1/2}$, where m is its mass and $c = C \, \rho_a \, A$. Writing $m = A \, d \, \rho$, where d is the typical linear dimension of the projectile and ρ its mass density, we obtain

$$v_t = \left(\frac{\rho \, g \, d}{\rho_a \, C} \right)^{1/2}. \tag{4.92}$$

The previous expression tells us that large, dense, streamlined projectiles (e.g., medicine balls) tend to have large terminal velocities, whereas small, rarefied, non-streamlined projectiles (e.g., feathers) tend to have small terminal velocities. Hence, the former type of projectile is relatively less affected by air resistance than the latter.

4.10 FRICTION

If a body slides over a rough surface then a frictional force generally develops that acts to impede the motion. Friction, when viewed at the microscopic level, is actually a very complicated phenomenon. Nevertheless, physicists and engineers have managed to develop a relatively simple empirical law of force that allows the effects of friction to be incorporated into their calculations. This law of force was first proposed by Leonardo da Vinci, and later extended by Charles Augustin de Coulomb (who is more famous for discovering the law of electrostatic attraction). According to this law, the frictional force exerted on a body sliding over a rough surface is proportional to the normal reaction, R_n, at that surface, the constant of proportionality depending on the nature of the surface. In other words,

$$f = \mu \, R_n, \tag{4.93}$$

where μ is termed the *coefficient of friction*. For ordinary surfaces, μ is generally of order unity.

Consider a block of mass m being dragged over a horizontal surface, whose coefficient of friction is μ, by a horizontal force F. See Figure 4.16. The weight, $W = m \, g$, of the block acts vertically downward, giving rise to a reaction, $R = m \, g$, acting vertically upward. The magnitude of the frictional force, f, that impedes the motion of the block is simply μ times the normal reaction, $R = m \, g$. Hence, $f = \mu \, m \, g$. The horizontal acceleration of the block is, therefore,

$$a = \frac{F - f}{m} = \frac{F}{m} - \mu \, g, \tag{4.94}$$

assuming that $F > f$. What happens if $F < f$; that is, if the applied force, F, is less than the frictional force, f? In this case, common sense suggests that the block simply remains at rest. Hence, $f = \mu \, m \, g$ is actually the maximum force that friction can generate in order to impede the motion of the block. If the applied force, F, is less than this maximum value

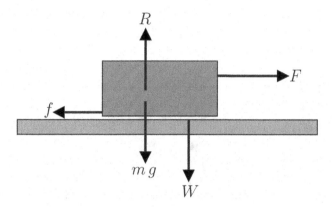

Figure 4.16 Block dragged over a rough surface

then the applied force is canceled out by an equal and opposite frictional force, and the block remains stationary. Only if the applied force exceeds the maximum frictional force does the block start to move.

Up to now, we have implicitly suggested that the coefficient of friction between an object and a surface is the same whether the object remains stationary or slides over the surface. In fact, this is generally not the case. Usually, the coefficient of friction when the object is stationary is slightly larger than the coefficient when the object is sliding. We call the former coefficient the *coefficient of static friction, μ_s*, whereas the latter coefficient is usually termed the *coefficient of kinetic (or dynamical) friction, μ_k*. The fact that $\mu_s > \mu_k$ simply implies that objects have a tendency to "stick" to rough surfaces when placed upon them. The force required to unstick a given object, and, thereby, set it in motion, is μ_s times the normal reaction at the surface. Once the object has been set into motion, the frictional force acting to impede this motion falls somewhat to μ_k times the normal reaction.

4.11 INCLINED PLANES

4.11.1 Smooth Planes

Consider a block of mass m sliding down a smooth frictionless incline that subtends an angle θ to the horizontal, as shown in Figure 4.17. The weight, $m\,g$, of the block is directed vertically downward. However, this force can be resolved into components $m\,g\,\cos\theta$, acting perpendicular (or normal) to the incline, and $m\,g\,\sin\theta$, acting parallel to the incline. Note that the reaction of the incline to the weight of the block acts normal to the incline, and only matches the normal component of the weight; that is, it is of magnitude $m\,g\,\cos\theta$. This is a general result; the reaction of any unyielding surface is always locally normal to that surface, directed outward (away from the surface), and matches the normal component of any inward force applied to the surface. The block is clearly in equilibrium in the direction normal to the incline, because the normal component of the block's weight is balanced by the reaction of the incline. However, the block is subject to an unbalanced force $m\,g\,\sin\theta$ in the direction parallel to the incline, and, therefore, accelerates down the slope. Applying Newton's second law to this problem (with the coordinates shown in the figure), we obtain

$$m\frac{d^2x}{dt^2} = m\,g\,\sin\theta, \tag{4.95}$$

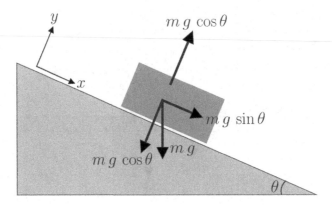

Figure 4.17 Block sliding down a smooth incline

which can be solved to give

$$x = x_0 + v_0\, t + \frac{1}{2}\, g\, \sin\theta\, t^{\,2}. \qquad (4.96)$$

In other words, the block accelerates down the slope with acceleration $g\sin\theta$. Note that this acceleration is less than the full acceleration due to gravity, g. In fact, if the incline is fairly gentle (i.e., if θ is small) then the acceleration of the block can be made much less than g. This was the technique used by Galileo in his pioneering studies of motion under gravity; by diluting the acceleration due to gravity, using inclined planes, he was able to obtain motion sufficiently slow for him to make accurate measurements using the crude time-keeping devices available in the 17th century.

Suppose, now, that the block is subject to an external force, F, that acts up the incline, parallel to its slope. If a is the block's upward acceleration, parallel to the slope, then its parallel equation of motion becomes

$$m\, a = F - m\, g\, \sin\theta. \qquad (4.97)$$

Note that the force, F, that must be applied to the block in order to push it very slowly up the incline is $W\sin\theta$, where $W = m\, g$ is the block's weight. This force is less that that required to lift the block directly; that is, W. In fact, if the incline is shallow then the force can be made very much less than W. Hence, as was well known to the ancients, a smooth inclined plane constitutes a very effective machine for lifting heavy weights.

4.11.2 Rough Planes

Suppose that the incline is rough. Let μ be its coefficient of friction. As before, the weight, $m\, g$, of the block can be resolved into components $m\, g\, \cos\theta$, acting normal to the incline, and $m\, g\, \sin\theta$, acting parallel to the incline. The reaction of the incline to the weight of the block acts normally outward from the incline, and is of magnitude $m\, g\, \cos\theta$.

Suppose that the block tries to slide down the incline. Parallel to the incline, the block is subject to the downward gravitational force $m\, g\, \sin\theta$, and the upward frictional force, f (which acts to prevent the block sliding down the incline). In order for the block to move, the magnitude of the former force must exceed the maximum value of the latter, which is μ time the magnitude of the normal reaction, or $f = \mu\, m\, g\, \cos\theta$. Hence, the condition for the weight of the block to overcome friction, and, thus, to cause the block to slide down the incline, is

$$m g \sin \theta > \mu m g \cos \theta, \tag{4.98}$$

or

$$\tan \theta > \mu. \tag{4.99}$$

In other words, if the slope of the incline exceeds a certain critical value, which depends on μ, then the block will start to slide. Incidentally, the previous formula suggests a fairly simple way of determining the coefficient of friction for a given object sliding over a particular surface. Simply tilt the surface gradually until the object just starts to move; the coefficient of friction is simply the tangent of the critical tilt angle (measured with respect to the horizontal).

4.12   FRAMES OF REFERENCE

As discussed in Section 1.4, it is a fundamental axiom of physics that physical laws possess objective reality. In other words, it is assumed that two independent observers, studying the same physical phenomenon, would eventually formulate equivalent laws of physics in order to account for their observations.

Now, two completely independent observers are likely to choose different systems of units with which to quantify physical measurements. However, as we have seen in Section 1.4, the dimensional consistency of valid laws of physics renders them invariant under transformation from one system of units to another.

Independent observers are also likely to choose different coordinate systems. For instance, the origins of their separate coordinate systems might differ, as well as the orientation of the various coordinate axes. However, as is clear from Chapter 3, if physical laws are expressed solely in terms of quantities that transform as scalars or vectors under rotation of the coordinate axes then such laws are independent of the orientation of these axes, or the location of the origin of the coordinate system. In particular, Newton's second law of motion,

$$m \mathbf{a} = \mathbf{f}, \tag{4.100}$$

is clearly invariant under shifts in the origin of the coordinate system, or changes in the orientation of the various coordinate axes. The reason for this is that the mass, m, is independent of the coordinate system, because measurements of mass do not involve measurements of distance. It follows that mass is a scalar quantity. The same is true of time. Furthermore, acceleration is directly linked to displacement, which is the prototype of all vectors. Hence, acceleration is a vector quantity. Finally, it is a matter of experience that forces transform as vectors under rotation of the coordinate axes. Hence, we conclude that Newton's second law of motion is equally valid in all coordinate systems. To sum up, valid laws of physics must consist solely of combinations of scalars and vectors, otherwise they would retain an unphysical dependence on the arbitrary details of the chosen coordinate system.

Up to now, we have implicitly assumed that all of our observers are stationary (i.e., they are all standing still on the surface of the Earth).[1] Let us, now, relax this assumption. Consider two observers, O and O', whose coordinate systems coincide momentarily at $t = 0$. Suppose that observer O is stationary (on the surface of the Earth), whereas observer O' moves (with respect to observer O) at the constant velocity $\mathbf{v}_0$. As illustrated in Figure 4.18, if $\mathbf{r}$ represents the displacement of some body P in the stationary observer's frame of

---

[1]Note that, for the moment, we are treating a frame of reference that is stationary on the surface of the Earth as inertial. Later on, in Chapter 12, we shall take into account the slightly non-inertial features of such a reference frame.

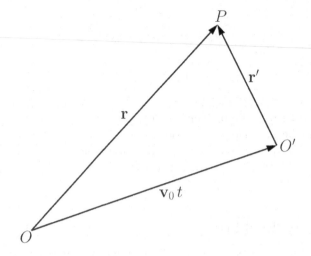

Figure 4.18   A moving observer

reference, at time t, then the corresponding displacement in the moving observer's frame of reference is simply

$$\mathbf{r}' = \mathbf{r} - \mathbf{v}_0 \, t. \tag{4.101}$$

The velocity of body P in the stationary observer's frame of reference is defined as

$$\mathbf{v} = \frac{d\mathbf{r}}{dt}, \tag{4.102}$$

whereas the corresponding velocity in the moving observer's frame of reference takes the form

$$\mathbf{v}' \equiv \frac{d\mathbf{r}'}{dt} = \mathbf{v} - \mathbf{v}_0. \tag{4.103}$$

Finally, the acceleration of body P in stationary observer's frame of reference is defined as

$$\mathbf{a} = \frac{d\mathbf{v}}{dt}, \tag{4.104}$$

whereas the corresponding acceleration in the moving observer's frame of reference takes the form

$$\mathbf{a}' \equiv \frac{d\mathbf{v}'}{dt} = \mathbf{a}. \tag{4.105}$$

Hence, the acceleration of body P is identical in both frames of reference.

It is clear that if observer O concludes that body P is moving with constant velocity, and, therefore, subject to zero net force, then observer O' will agree with this conclusion. Furthermore, if the observer O concludes that body P is accelerating, and, therefore, subject to a force $\mathbf{a}/m$, then observer O' will remain in agreement. It follows that Newton's laws of motion are equally valid in the frames of reference of the moving and the stationary observer. Such frames are termed *inertial frames of reference*. There are infinitely many inertial frames of reference—within which Newton's laws of motion are equally valid—all moving with constant velocities with respect to one another. Consequently, there is no universal standard of rest in physics. Observer O might claim to be at rest compared to observer O', and vice versa. However, both points of view are equally valid. Moreover, there is absolutely no physical experiment that observer O could perform in order to demonstrate that he/she is at rest while observer O' is moving. This, in essence, is the principle of the *special theory of relativity*, first formulated by Albert Einstein in 1905.

4.13 EXERCISES

4.1 A balloon whose net mass is M is falling with acceleration f. What amount of ballast must be let loose in order that the balloon may have an upward acceleration of f? (Assume that loosing the ballast does not affect the buoyancy of the balloon.) [Ans: $2 f M/(f+g)$.] [From Lamb 1942.]

4.2 Two equal masses are suspended by five cables, as shown in Figure 4.8. Given that $\theta = 50°$ and $T_1 = 50$ N, determine the values of m and T_2. [Ans: 6.07 kg and 77.79 N.]

4.3 A uniform chain of weight W is suspended from two supports of equal height.

(a) Show that the maximum tension in the chain is

$$T = \sqrt{H^2 + W^2/4},$$

where H is the horizontal tension.

(b) Demonstrate that

$$\frac{H}{W} = \frac{L}{8d} - \frac{d}{2L},$$

where L is the length of the chain, and d is its sag. Hence, deduce that the tension in a chain of fixed length and weight is minimized when $d = L/2$. Show that, when in this optimal configuration, $H = W/2$, the two ends of the chain subtend an angle of $45°$ with the horizontal, and the tension at the two ends is $T = W/\sqrt{2}$.

4.4 A uniform cable of weight per unit length 7 N/m is suspended from two vertical poles of equal height that are a horizontal distance 30 m apart. The horizontal tension in the cable is 800 N.

(a) What is the sag in the cable? [Ans: 0.986 m.]

(b) What is the cable tension at the two poles? [Ans: 806.9 N.]

4.5 A horizontal load of 2000 N/m is supported by a light cable suspended from poles at the same level and 20 m apart. The maximum tension in the cable is 140 kN.

(a) What is the sag in the cable? [Ans: 0.72 m.]

(b) What is the required length of cable? [Ans: 20.07 m.]

4.6 A block of mass 5 kg lies on a frictionless incline that subtends an angle of $25°$ with the horizontal, and is subject to a horizontal force of 27 N that acts to push it up the incline. What is the acceleration of the block? [Ans: 0.748 m/s^2.]

4.7 A block of mass m_1 is being pulled up a smooth incline that subtends an angle θ with the horizontal by a cable that runs parallel to the incline. The cable is draped over a light frictionless pulley at the top of the slope. The other end of the cable is attached to a block of mass m_1 that is suspended in a vertical shaft. Show that the tension in the cable and the downward acceleration of the second mass are

$$T = \frac{m_1 m_2 (1 + \sin \theta) g}{m_1 + m_2},$$

$$a = \frac{(m_2 - m_1 \sin \theta) g}{m_1 + m_2},$$

respectively.

4.8 Consider a flexible chain of length l and uniform mass per unit length μ. At time $t = 0$, the chain is at rest on a smooth horizontal table, with a section of length c overhanging the edge of the table. Let $x(t)$ be the length of the section of the chain overhanging the table at time t.

   (a) How does $x(t)$ vary in time? [Ans: $x = c \cosh(\sqrt{g/l}\,t)$.]

   (b) What is the instantaneous speed of the chain as its last link slides off the table? [Ans: $\sqrt{g/l}\,\sqrt{l^2 - c^2}$.]

   (c) What is the instantaneous acceleration of the chain as its last link slides off the table? [Ans: g.]

4.9 Consider the table-pulley system shown in Figure 4.14. Suppose that the coefficient of friction of the table is μ.

   (a) What is the acceleration of the system? [Ans: $(m_2 - \mu m_1) g/(m_1 + m_2)$ if $\mu < m_2/m_1$, and 0 otherwise.]

   (b) What is the tension in the cable? [Ans: $m_1 m_2 (1+\mu) g/(m_1+m_2)$ if $\mu < m_2/m_1$, and $m_2 g$ otherwise.]

4.10 In an Atwood machine, two masses of 40 kg and 35 kg, respectively, are attached by a cable that passes over a frictionless pulley. If the masses start from rest, find the distance covered by either mass in 6 s. [Ans: 11.3 m.]

4.11 If a train of mass M is subject to a retarding force $M (a + b v^2)$, show that if the engines are shut off when the speed is v_0 then the train will come to rest in a time

$$\frac{1}{\sqrt{a b}} \tan^{-1} \left(\sqrt{\frac{b}{a}} v_0 \right),$$

after traveling a distance

$$\frac{1}{2 b} \ln \left(1 + \frac{b v_0^2}{a} \right).$$

4.12 A block of mass m slides along a horizontal surface that is lubricated with heavy oil such that the block suffers a viscous retarding force of the form

$$F = -c v^n,$$

where $c > 0$ is a constant, and v is the block's instantaneous velocity. The block's initial (at $t = 0$) speed and displacement are v_0 and 0, respectively.

   (a) Demonstrate that

$$v^{1-n} - v_0^{1-n} = -(1 - n) \frac{c t}{m}.$$

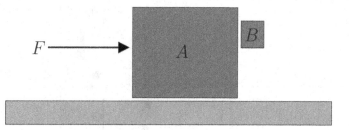

Figure 4.19 Figure for Exercise 4.17

(b) Demonstrate that

$$v^{2-n} - v_0^{2-n} = -(2-n)\frac{c\,x}{m}.,\text{ where }x\text{ is the displacement of block from}$$

its initial position.

(c) Show that for $n = 1/2$ the block does not travel further than $2\,m\,v_0^{3/2}/(3\,c)$.

[Modified from Fowles & Cassiday 2005.]

4.13 A particle is projected vertically upward in a constant gravitational field with an initial speed v_0. Show that if there is a retarding force proportional to the square of the speed then the speed of the particle when it returns to the initial position is

$$\frac{v_0\,v_t}{\sqrt{v_0^2 + v_t^2}},$$

where v_t is the terminal speed. [From Thornton & Marion 2004.]

4.14 A particle of mass m moves (in one dimension) in a medium under the influence of a retarding force of the form $m\,k\,(v^3 + a^2\,v)$, where v is the particle speed, and k and a are positive constants. Show that for any value of the initial speed the particle will never move a distance greater than $\pi/(2\,k\,a)$, and will only come to rest as $t \to \infty$. [From Thornton & Marion 2004.]

4.15 A block B rests on another block A that is being pulled along a smooth horizontal surface by a horizontal force P. If the coefficient of friction between the two blocks is μ, determine the maximum acceleration of the system before slippage occurs between A and B. [Ans: $\mu\,g$.]

4.16 Two blocks of mass m_1 and m_2 are connected together via a light rigid strut, and are placed on a rough slope whose inclination to the horizontal is θ. Let the block whose mass is m_1 be higher up the slope than the block whose mass is m_2. Let μ_1 and μ_2 be the coefficients of friction between the blocks whose masses are m_1 and m_2 and the slope, respectively. Suppose that the strut is parallel to the slope.

(a) What is the critical value of θ that must be exceeded in order for the two blocks to slide down the slope? [Ans: $\tan^{-1}[(\mu_1\,m_1 + \mu_2\,m_2)/(m_1 + m_2)]$.]

(b) What is the tension in the strut when θ takes this critical value? [Ans: $(\mu_1 - \mu_2)\,m_1\,m_2\,g/[(m_1 + m_2)^2 + (\mu_1\,m_1 + \mu_2\,m_2)^2]^{1/2}$.]

4.17 Consider Figure 4.19. The mass of block A is 75 kg, and the mass of block B is 15 kg. The coefficient of static friction between the two blocks is $\mu = 0.45$. The horizontal surface is frictionless. What minimum force, F, must be exerted on block A in order to prevent block B from falling? [Ans: 1.96×10^3 N.]

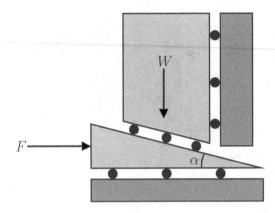

Figure 4.20 Figure for Exercise 4.18

4.18 Figure 4.20 shows how a weight can be slowly raised by an ideal frictionless wedge.

(a) What is the relationship between the load, W, the applied force, F, and the wedge angle, α? [Ans: $F = W \tan \alpha$.]

(b) What is the force transmitted by the vertical row of balls? [Ans: F.]

(c) What is the force transmitted by the horizontal row of balls? [Ans: W.]

(d) What is the force transmitted by the inclined row of balls? [Ans: $\sqrt{F^2 + W^2}$.]

[From Den Hartog 1961.]

4.19 A wedge consists of two inclined planes that meet at an acute angle θ, and is used to split a log. If μ is the coefficient of friction between the wedge and the log, what is the condition that must be satisfied for the wedge not to be squeezed out of the log when no external force is applied to it. [Ans: $\tan(\theta/2) < \mu$.] [From Den Hartog 1961.]

CHAPTER 5

Conservation of Energy

5.1 INTRODUCTION

Nowadays, the conservation of energy is undoubtedly the single most important idea in physics. Strangely enough, although the basic idea of energy conservation was familiar to scientists from the time of Newton onward, this crucial concept only moved to center stage in physics in about 1850 (i.e., when scientists first realized that heat was a form of energy).

According to the ideas of modern physics, **energy** is the fundamental substance that makes up all things in the universe. Energy can take many different forms; for instance, potential energy, kinetic energy, electrical energy, thermal energy, chemical energy, nuclear energy, etcetera. In fact, everything that we observe in the world around us represents one of the multitudinous manifestations of energy. There exist processes in the universe that transform energy from one form into another; for instance, mechanical processes (which are the focus of this book), thermal processes, electrical processes, nuclear processes, etcetera. However, all of these processes leave the total amount of energy in the universe invariant. In other words, whenever, and however, energy is transformed from one form into another, it is always conserved. For a closed system (i.e., a system that does not exchange energy with the rest of the universe), the previous law of universal energy conservation implies that the total energy of the system in question must remain constant in time.

5.2 ENERGY CONSERVATION DURING FREE-FALL

Consider a mass, m, that is falling vertically under the influence of gravity. We already know how to analyze the motion of such a mass. Let us employ this knowledge to search for an expression for the conserved energy during this process. (Note that this is clearly an example of a closed system, involving only the mass and the gravitational field.) The physics of free-fall under gravity is summarized by the three equations (2.16)–(2.18). Let us examine the last of these equations;

$$v^2 = v_0^2 - 2\,g\,s. \tag{5.1}$$

Suppose that the mass falls from height (measured with respect to sea level) h_1 to h_2, that its initial velocity is v_1, and that its final velocity is v_2. It follows that the net vertical displacement of the mass is $s = h_2 - h_1$. Moreover, $v_0 = v_1$ and $v = v_2$. Hence, the previous expression can be rearranged to give

$$\frac{1}{2}\,m\,v_1^2 + m\,g\,h_1 = \frac{1}{2}\,m\,v_2^2 + m\,g\,h_2. \tag{5.2}$$

DOI: 10.1201/9781003198642-5

The previous equation clearly represents a conservation law, of some description, because the left-hand side only contains quantities evaluated at the initial height, whereas the right-hand side only contains quantities evaluated at the final height. In order to clarify the meaning of Equation (5.2), let us define the *kinetic energy* of the mass,

$$K = \frac{1}{2} m v^2,\qquad(5.3)$$

and the *gravitational potential energy* of the mass,

$$U = m g h.\qquad(5.4)$$

Here, we are assuming that energy possesses units of newton-meters (i.e., joules). Note that kinetic energy represents energy the mass possesses by virtue of its motion. Likewise, potential energy represents energy the mass possesses by virtue of its position. It follows that Equation (5.2) can be written

$$E = K + U = \text{constant}.\qquad(5.5)$$

Here, E is the total energy of the mass; that is, the sum of its kinetic and potential energies. It is clear that E is a conserved quantity; that is, although the kinetic and potential energies of the mass vary as it falls, its total energy remains the same.

Incidentally, the expressions (5.3) and (5.4) for kinetic and gravitational potential energy, respectively, are quite general, and do not just apply to free-fall under gravity.

Generally speaking, there are many different paths to the same result in physics. We have already analyzed free-fall under gravity using Newton's laws of motion. (See Section 2.8.) However, it is illuminating to re-examine this problem from the point of view of energy conservation. Suppose that a mass, m, is dropped from rest, and falls a distance h. What is the final speed, u, of the mass? According to Equation (5.2), if energy is conserved then

$$\Delta K = -\Delta U;\qquad(5.6)$$

that is, any increase in the kinetic energy of the mass must be offset by a corresponding decrease in its potential energy. Now, the change in potential energy of the mass is simply $\Delta U = m g s = -m g h$, where $s = -h$ is its net vertical displacement. The change in kinetic energy is simply $\Delta K = (1/2) m u^2$. This follows because the initial kinetic energy of the mass is zero (because it is initially at rest). Hence, the previous expression yields

$$\frac{1}{2} m u^2 = m g h,\qquad(5.7)$$

or

$$u = \sqrt{2 g h}.\qquad(5.8)$$

[See Equation (2.20).]

Suppose that the same mass is thrown upward with initial velocity u. What is the maximum height, h, to which it rises? It is clear from Equation (5.4) that as the mass rises its potential energy increases. It, therefore, follows from energy conservation that its kinetic energy must decrease with height. Note, however, from Equation (5.3), that kinetic energy can never be negative (because it is the product of the two positive definite quantities, m and $v^2/2$). Hence, once the mass has risen to a height h which is such that its kinetic energy is reduced to zero, it can rise no further, and must, presumably, start to fall. The change in potential energy of the mass in moving from its initial height to its maximum height is $m g h$. The corresponding change in kinetic energy is $-(1/2) m u^2$; because $(1/2) m u^2$ is the initial kinetic energy, and the final kinetic energy is zero. It follows from Equation (5.6) that $-(1/2)\, m\, u^2 = -m\, g\, h$, which can be rearranged to give

$$h = \frac{u^2}{2\, g}.$$ (5.9)

[See Equation (2.21).]

It should be noted that the idea of energy conservation—although extremely useful—is not a replacement for Newton's laws of motion. For instance, in the previous example, there is no way in which we can deduce how long it takes the mass to rise to its maximum height from energy conservation alone; this information can only come from the direct application of Newton's laws.

5.3 WORK

We have seen that when a mass free-falls under the influence of gravity some of its kinetic energy is transformed into potential energy. Let us now investigate, in detail, how this transformation is effected. The mass falls because it is subject to a downward gravitational force of magnitude $m\, g$. It stands to reason, therefore, that the transformation of kinetic into potential energy is a direct consequence of the action of this force.

This is, perhaps, an appropriate point at which to note that the concept of gravitational potential energy—although extremely useful—is, strictly speaking, fictitious. To be more exact, the potential energy of a body is not an intrinsic property of that body (unlike its kinetic energy). In fact, the gravitational potential energy of a given body is stored in the gravitational field that surrounds it. Thus, when the body rises, and its potential energy consequently increases by an amount ΔU, in reality, it is the energy of the gravitational field surrounding the body that increases by this amount. Of course, the increase in energy of the gravitational field is offset by a corresponding decrease in the body's kinetic energy. Thus, when we speak of a body's kinetic energy being transformed into potential energy, we are really talking about a flow of energy from the body to the surrounding gravitational field. This energy flow is mediated by the gravitational force exerted by the field on the body in question.

Incidentally, according to Einstein's *general theory of relativity* (1915), the gravitational field of a mass consists of the local distortion that the mass induces in the fabric of space-time. Fortunately, however, we do not need to understand general relativity in order to talk about gravitational fields or gravitational potential energy. All we need to know is that a gravitational field stores energy without loss; that is, if a given mass rises a certain distance, and, thereby, gives up a certain amount of energy to the surrounding gravitational field, then that field will return this energy to the mass without loss if the mass falls by the same distance. In physics, we term such a field a conservative field.

Suppose that a mass m falls a distance $-\Delta h$, where h measures height about sea level. During this process, the energy of the gravitational field decreases by a certain amount (i.e., the fictitious potential energy of the mass decreases by a certain amount), and the body's kinetic energy increases by a corresponding amount. This transfer of energy, from the field to the mass, is, presumably, mediated by the gravitational force $f = -m\, g$ (the minus sign indicates that the force is directed downward) acting on the mass. In fact, given that $U = m\, g\, h$, it follows from Equation (5.6) that

$$\Delta K = f\, \Delta h.$$ (5.10)

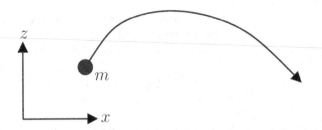

Figure 5.1   Coordinate system for two-dimensional motion under gravity

In other words, the amount of energy transferred to the mass (i.e., the increase in the mass's kinetic energy) is equal to the product of the force acting on the mass and the distance moved by the mass in the direction of that force.

In physics, we generally refer to the amount of energy transferred to a body, when a force acts upon it, as the amount of *work*, W, performed by that force on the body in question. It follows from Equation (5.10) that when a gravitational force f acts on a body, causing it to displace a distance x in the direction of that force, then the net work done on the body is

$$W = f\,x. \tag{5.11}$$

It turns out that this equation is quite general, and does not just apply to gravitational forces. If W is positive then energy is transferred to the body, and its intrinsic energy consequently increases by an amount W. This situation occurs whenever a body moves in the same direction as the force acting upon it. Likewise, if W is negative then energy is transferred from the body, and its intrinsic energy consequently decreases by an amount $|W|$. This situation occurs whenever a body moves in the opposite direction to the force acting upon it. Because an amount of work is equivalent to a transfer of energy, the mks unit of work is the same as the mks unit of energy; namely, the joule.

In deriving equation (5.11), we have made two assumptions that are not universally valid. First, we have assumed that the motion of the body upon which the force acts is both one-dimensional and parallel to the line of action of the force. Second, we have assumed that the force does not vary with position. Let us attempt to relax these two assumptions, so as to obtain an expression for the work, W, done by a general force, $\mathbf{f}$.

Let us start by relaxing the first assumption. Suppose, for the sake of argument, that we have a mass, m, that moves under gravity in two dimensions. Let us adopt the coordinate system shown in Figure 5.1, with z representing vertical distance, and x representing horizontal distance. The vector acceleration of the mass is simply $\mathbf{a} = (0, -g)$. Here, we are neglecting the redundant y-component of the acceleration, for the sake of simplicity. The physics of motion under gravity in more than one dimension is summarized by the three equations (3.62)–(3.64). Let us examine the last of these equations:

$$v^2 = v_0^2 + 2\,\mathbf{a} \cdot \mathbf{s}. \tag{5.12}$$

Here, v_0 is the speed at $t = 0$, v is the speed at $t = t$, and $\mathbf{s} = (\Delta x, \ \Delta z)$ is the net displacement of the mass during this time interval. Recalling the definition of a scalar product [i.e., $\mathbf{a} \cdot \mathbf{b} = (a_x\,b_x + a_y\,b_y + a_z\,b_z)$], the previous equation can be rearranged to give

$$\frac{1}{2}\,m\,v^2 - \frac{1}{2}\,m\,v_0^2 = -m\,g\,\Delta z. \tag{5.13}$$

Because the left-hand side of the previous expression is equal to the increase in the kinetic energy of the mass between times $t = 0$ and $t = t$, the right-hand side must equal the

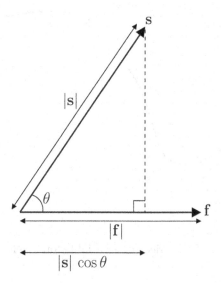

Figure 5.2 Definition of work.

decrease in the mass's potential energy during the same time interval. Hence, we arrive at the following expression for the gravitational potential energy of the mass;

$$U = m\,g\,z. \tag{5.14}$$

Of course, this expression is entirely equivalent to our previous expression for gravitational potential energy, Equation (5.4). The previous expression merely makes manifest a point that should have been obvious anyway; namely, that the gravitational potential energy of a mass only depends on its height above the sea level, and is quite independent of its horizontal displacement.

Let us now try to relate the flow of energy between the gravitational field and the mass to the action of the gravitational force, $\mathbf{f} = (0, -m\,g)$. Equation (5.13) can be rewritten

$$\Delta K = W = \mathbf{f} \cdot \mathbf{s}. \tag{5.15}$$

In other words, the work, W, done by the force $\mathbf{f}$ is equal to the scalar product of $\mathbf{f}$ and the vector displacement, $\mathbf{s}$, of the body upon which the force acts. It turns out that this result is quite general, and does not just apply to gravitational forces.

Figure 5.2 is a visualization of the definition (5.15). The work, W, performed by a force $\mathbf{f}$ when the object upon which it acts is subject to a displacement $\mathbf{s}$ is

$$W = |\mathbf{f}|\,|\mathbf{s}|\,\cos\theta. \tag{5.16}$$

where θ is the angle subtended between the directions of $\mathbf{f}$ and $\mathbf{s}$. In other words, the work performed is the product of the magnitude of the force, $|\mathbf{f}|$, and the displacement of the object in the direction of that force, $|\mathbf{s}|\,\cos\theta$. It follows that any component of the displacement in a direction perpendicular to the force generates zero work. Moreover, if the displacement is entirely perpendicular to the direction of the force (i.e., if $\theta = 90°$) then no work is performed, irrespective of the nature of the force. As before, if the displacement

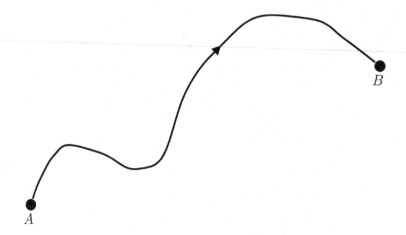

Figure 5.3 Possible trajectory of an object in a variable force-field

has a component in the same direction as the force (i.e., if $\theta < 90°$) then positive work is performed. Likewise, if the displacement has a component in the opposite direction to the force (i.e., if $\theta > 90°$) then negative work is performed.

Suppose, now, that an object is subject to a force, $\mathbf{f}$, that varies with position. What is the total work done by the force when the object moves along some general trajectory in space between points A and B (say)? See Figure 5.3. One way in which we could approach this problem would be to approximate the trajectory as a series of N straight-line segments, as shown in Figure 5.4. Suppose that the vector displacement of the ith segment is $\mathbf{\Delta r}_i$. Suppose, further, that N is sufficiently large that the force $\mathbf{f}$ does not vary much along each segment. In fact, let the average force along the ith segment be $\mathbf{f}_i$. We shall assume that formula (5.15)—which is valid for constant forces and straight-line displacements—holds good for each segment. It follows that the net work done on the body, as it moves from point A to point B, is approximately

$$W \simeq \sum_{i=1,N} \mathbf{f}_i \cdot \mathbf{\Delta r}_i. \tag{5.17}$$

We can always improve the level of our approximation by increasing the number, N, of the straight-line segments that we use to approximate the body's trajectory between points A and B. In fact, if we take the limit $N \to \infty$ then the previous expression becomes exact:

$$W = \lim_{N \to \infty} \sum_{i=1,N} \mathbf{f}_i \cdot \mathbf{\Delta r}_i = \int_A^B \mathbf{f}(\mathbf{r}) \cdot d\mathbf{r}. \tag{5.18}$$

Here, $\mathbf{r}$ measures vector displacement from the origin of our coordinate system, and the mathematical construct $\int_A^B \mathbf{f}(\mathbf{r}) \cdot d\mathbf{r}$ is termed a *line integral*.

The meaning of Equation (5.18) becomes a lot clearer if we restrict our attention to one-dimensional motion. Suppose, therefore, that an object moves in one dimension, with displacement x, and is subject to a varying force $f(x)$ (directed along the x-axis). What is the work done by this force when the object moves from x_A to x_B? A straightforward application of Equation (5.18) [with $\mathbf{f} = (f, 0, 0)$ and $d\mathbf{r} = (dx, 0, 0)$] yields

$$W = \int_{x_A}^{x_B} f(x)\, dx. \tag{5.19}$$

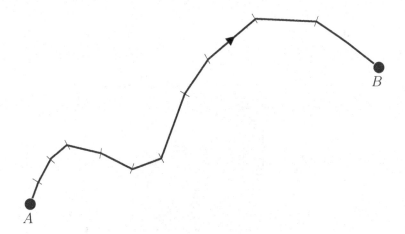

Figure 5.4 Approximation to the previous trajectory using straight-line segments

In other words, the net work done by the force as the object moves from displacement x_A to x_B is simply the area under the $f(x)$ curve between these two points, as illustrated in Figure 5.5.

Let us continue our discussion by re-deriving the so-called *work-energy theorem*, Equation (5.15), in one dimension, allowing for a non-constant force. According to Newton's second law of motion,

$$f = m\,\frac{d^2x}{dt^2}.$$
(5.20)

Combining Equations (5.19) and (5.20), we obtain

$$W = \int_{x_A}^{x_B} m\,\frac{d^2x}{dt^2}\,dx = \int_{t_A}^{t_B} m\,\frac{d^2x}{dt^2}\frac{dx}{dt}\,dt = \int_{t_A}^{t_B} \frac{d}{dt}\left[\frac{m}{2}\left(\frac{dx}{dt}\right)^2\right]dt,$$
(5.21)

where $x(t_A) = x_A$ and $x(t_B) = x_B$. It follows that

$$W = \frac{1}{2}\,m\,v_B^2 - \frac{1}{2}\,m\,v_A^2 = \Delta K,$$
(5.22)

where $v_A = (dx/dt)_{t_A}$ and $v_B = (dx/dt)_{t_B}$. Thus, the net work performed on a body by a non-uniform force, as it moves from point A to point B, is equal to the net increase in that body's kinetic energy between these two points. This result is completely general (at least, for conservative force-fields), and does not just apply to one-dimensional motion.

Suppose, finally, that an object is subject to more than one force. How do we calculate the net work, W, performed by all of these forces as the object moves from point A to point B? One approach would be to calculate the work done by each force, taken in isolation, and then to sum the results. In other words, defining

$$W_i = \int_A^B \mathbf{f}_i(\mathbf{r}) \cdot d\mathbf{r}$$
(5.23)

as the work done by the ith force, the net work is given by

$$W = \sum_i W_i.$$
(5.24)

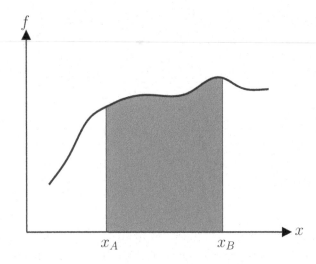

Figure 5.5   Work performed by a one-dimensional force.

An alternative approach would be to take the vector sum of all the forces to find the resultant force,

$$\mathbf{f} = \sum_i \mathbf{f}_i, \tag{5.25}$$

and then to calculate the work done by the resultant force:

$$W = \int_A^B \mathbf{f}(\mathbf{r}) \cdot d\mathbf{r}. \tag{5.26}$$

It should, hopefully, be clear that these two approaches are entirely equivalent.

5.4   CONSERVATIVE AND NON-CONSERVATIVE FORCE-FIELDS

Suppose that a non-uniform force-field, $\mathbf{f}(\mathbf{r})$, acts upon an object that moves along a curved trajectory, labeled path 1, from point A to point B. See Figure 5.6. As we have seen, the work, W_1, performed by the force-field on the object can be written as a line integral along this trajectory:

$$W_1 = \int_{A \to B:\,\text{path 1}} \mathbf{f} \cdot d\mathbf{r}. \tag{5.27}$$

Suppose that the same object moves along a different trajectory, labeled path 2, between the same two points. In this case, the work, W_2, performed by the force-field is

$$W_2 = \int_{A \to B:\,\text{path 2}} \mathbf{f} \cdot d\mathbf{r}. \tag{5.28}$$

Basically, there are two possibilities. Firstly, the line integrals (5.27) and (5.28) might depend on the end points, A and B, but not on the path taken between them, in which case $W_1 = W_2$. Secondly, the line integrals (5.27) and (5.28) might depend both on the end points, A and B, and the path taken between them, in which case $W_1 \neq W_2$ (in general). The first possibility corresponds to what physicists term a *conservative* force-field, whereas the second possibility corresponds to a *non-conservative* force-field.

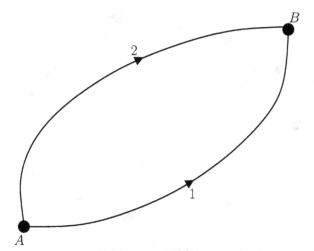

Figure 5.6 Two alternative paths between points A and B

What is the physical distinction between a conservative and a non-conservative force-field? The easiest way of answering this question is to slightly modify the problem just discussed. Suppose, now, that the object moves from point A to point B along path 1, and then from point B back to point A along path 2. What is the total work done on the object by the force-field as it executes this closed circuit? Incidentally, one fact that should be clear from the definition of a line integral is that if we simply reverse the path of a given integral then the value of that integral picks up a minus sign; in other words,

$$\int_A^B \mathbf{f} \cdot d\mathbf{r} = -\int_B^A \mathbf{f} \cdot d\mathbf{r}, \tag{5.29}$$

where it is understood that both the previous integrals are taken in opposite directions along the same path. Recall that conventional one-dimensional integrals obey an analogous rule; that is, if we swap the limits of integration then the integral picks up a minus sign. It follows that the total work done on the object as it executes the circuit is simply

$$\Delta W = W_1 - W_2, \tag{5.30}$$

where W_1 and W_2 are defined in Equations (5.27) and (5.28), respectively. There is a minus sign in front of W_2 because we are moving from point B to point A, instead of the other way around. For the case of a conservative field, we have $W_1 = W_2$. Hence, we conclude that

$$\Delta W = 0. \tag{5.31}$$

In other words, the net work done by a conservative field on an object taken around a closed loop is zero. This is just another way of saying that a conservative field stores energy without loss; that is, if an object gives up a certain amount of energy to a conservative field in traveling from point A to point B then the field returns this energy to the object, without loss, when it travels back to point B. For the case of a non-conservative field, $W_1 \neq W_2$. Hence, we conclude that

$$\Delta W \neq 0. \tag{5.32}$$

In other words, the net work done by a non-conservative field on an object taken around a closed loop is non-zero. In practice, the net work is invariably negative. This is just another

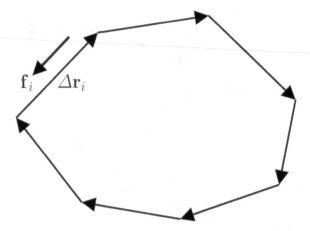

Figure 5.7 Closed circuit over a rough horizontal surface

way of saying that a non-conservative field dissipates energy; that is, if an object gives up a certain amount of energy to a non-conservative field in traveling from point A to point B then the field only returns part, or, perhaps, none, of this energy to the object when it travels back to point B. The remainder is usually dissipated as heat.

What are typical examples of conservative and non-conservative fields? A gravitational field is probably the most well-known example of a conservative field. A typical example of a non-conservative field might consist of an object moving over a rough horizontal surface. Suppose, for the sake of simplicity, that the object executes a closed circuit on the surface that is made up entirely of straight-line segments, as shown in Figure 5.7. Let $\Delta \mathbf{r}_i$ represent the vector displacement of the ith leg of this circuit. Suppose that the frictional force acting on the object as it executes this leg is $\mathbf{f}_i$. One thing that we know about a frictional force is that it is always directed in the opposite direction to the instantaneous direction of motion of the object upon which it acts. Hence, $\mathbf{f}_i \propto -\Delta \mathbf{r}_i$. It follows that $\mathbf{f}_i \cdot \Delta \mathbf{r}_i = -|\mathbf{f}_i|\,|\Delta \mathbf{r}_i|$. Thus, the net work performed by the frictional force on the object, as it executes the circuit, is given by

$$\Delta W = \sum_i \mathbf{f}_i \cdot \Delta \mathbf{r}_i = -\sum_i |\mathbf{f}_i|\,|\Delta \mathbf{r}_i| < 0. \tag{5.33}$$

The fact that the net work is negative indicates that the frictional force continually drains energy from the object as it moves over the surface. This energy is actually dissipated as heat (it is well known that if two rough surfaces are rubbed together then heat is generated; this is how mankind first made fire), and is, therefore, lost to the system. (Generally speaking, the laws of thermodynamics forbid energy that has been converted into heat from being directly converted back into its original form.) Hence, friction is an example of a non-conservative force, because it dissipates energy rather than storing it.

5.5   POTENTIAL ENERGY

Consider a body moving in a conservative force-field, $\mathbf{f}(\mathbf{r})$. Let us arbitrarily pick some point O in this field. We can define a function $U(\mathbf{r})$ that possesses a unique value at every point in the field. The value of this function associated with some general point R is simply

$$U(R) = -\int_O^R \mathbf{f} \cdot d\mathbf{r}. \tag{5.34}$$

In other words, $U(R)$ is just the energy transferred to the field (i.e., minus the work done by the field) when the body moves from point O to point R. Of course, the value of U at point O is zero; that is $U(O) = 0$. Note that the previous definition uniquely specifies $U(R)$, because the work done when a body moves between two points in a conservative force-field is independent of the path taken between these points. Furthermore, the previous definition would make no sense in a non-conservative field, because the work done when a body moves between two points in such a field is dependent on the chosen path; hence, $U(R)$ would have an infinite number of different values corresponding to the infinite number of different paths that the body could take between points O and R.

According to the work-energy theorem,

$$\Delta K = \int_O^R \mathbf{f} \cdot d\mathbf{r}. \tag{5.35}$$

In other words, the net change in the kinetic energy of the body, as it moves from point O to point R, is equal to the work done on the body by the force-field during this process. However, comparing with Equation (5.34), we can see that

$$\Delta K = U(O) - U(R) = -\Delta U. \tag{5.36}$$

In other words, the increase in the kinetic energy of the body, as it moves from point O to point R, is equal to the decrease in the function U evaluated between these same two points. Another way of putting this is

$$E = K + U = \text{constant}; \tag{5.37}$$

that is, the sum of the kinetic energy and the function U remains constant as the body moves around in the force-field. It should be clear, by now, that the function U represents some form of potential energy.

The previous discussion leads to the following important conclusions. First, it should be possible to associate a potential energy (i.e., an energy a body possesses by virtue of its position) with any conservative force-field. Second, any force-field for which we can define a potential energy must necessarily be conservative. For instance, the existence of gravitational potential energy proves that gravitational fields are conservative. Third, the concept of potential energy is meaningless in a non-conservative force-field (because the potential energy at a given point cannot be uniquely defined). Fourth, potential energy is only defined to within an arbitrary additive constant. In other words, the point in space at which we set the potential energy to zero can be chosen at will. This implies that only differences in potential energies between different points in space have any physical significance. For instance, we have seen that the definition of gravitational potential energy is $U = m\,g\,z$, where z represents height above sea level. However, we could just as well write $U = m\,g\,(z - z_0)$, where z_0 is the height of some arbitrarily chosen reference point (e.g., the top of Mount Everest, or the bottom of the Dead Sea). Fifth, the difference in potential energy between two points represents the net energy transferred to the associated force-field when a body moves between these two points. In other words, potential energy is not, strictly speaking, a property of the body; instead, it is a property of the force-field within which the body moves.

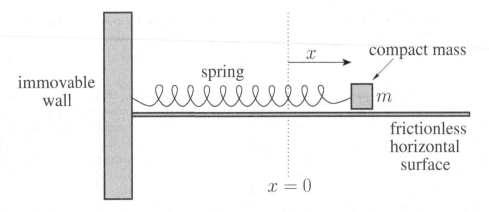

Figure 5.8 Mass on a spring. (Reproduced from Fitzpatrick 2019. Courtesy of Taylor & Francis.)

5.6 HOOKE'S LAW

Consider a mass, m, that slides over a horizontal frictionless surface. Suppose that the mass is attached to a light horizontal spring whose other end is anchored to an immovable object. See Figure 5.8. Let x be the extension of the spring; that is, the difference between the spring's actual length and its unstretched length. Obviously, x can also be used as a coordinate to determine the horizontal displacement of the mass. According to Hooke's law, the force, f, that the spring exerts on the mass is directly proportional to its extension, and always acts to reduce this extension. Hence, we can write

$$f = -k\,x, \tag{5.38}$$

where the positive quantity k is called the *force constant* of the spring. Note that the minus sign in the previous equation ensures that the force always acts to reduce the spring's extension; that is, if the extension is positive then the force acts to the left, so as to shorten the spring.

According to Equation (5.19), the work performed by the spring force on the mass as it moves from displacement x_A to x_B is

$$W = \int_{x_A}^{x_B} f(x)\,dx = -k \int_{x_A}^{x_B} x\,dx = -\frac{1}{2}\,k\,x_B^2 + \frac{1}{2}\,k\,x_A^2. \tag{5.39}$$

Note that the right-hand side of the previous expression consists of the difference between two factors; the first only depends on the final state of the mass, whereas the second only depends on its initial state. This is a sure sign that it is possible to associate a potential energy with the spring force. Equation (5.34), which is the basic definition of potential energy, yields

$$U(x_B) - U(x_A) = -\int_{x_A}^{x_B} f(x)\,dx = \frac{1}{2}\,k\,x_B^2 - \frac{1}{2}\,k\,x_A^2. \tag{5.40}$$

Hence, the potential energy of the mass takes the form

$$U(x) = \frac{1}{2}\,k\,x^2. \tag{5.41}$$

Note that the previous potential energy actually represents energy stored by the spring, in the form of mechanical stresses, when it is either stretched or compressed. Incidentally, this energy must be stored without loss, otherwise the concept of potential energy would be meaningless. It follows that the spring force is another example of a conservative force.

It is reasonable to suppose that the form of the spring potential energy is somehow related to the form of the spring force. Let us now explicitly investigate this relationship. If we let $x_B \rightarrow x$ and $x_A \rightarrow 0$ then Equation (5.40) gives

$$U(x) = -\int_0^x f(x')\,dx'. \tag{5.42}$$

We can differentiate this expression to obtain

$$f = -\frac{dU}{dx}. \tag{5.43}$$

Thus, in one dimension, a conservative force is equal to minus the derivative (with respect to displacement) of its associated potential energy. This is a quite general result. For the case of a spring force; $U = (1/2)\,k\,x^2$, so $f = -dU/dx = -k\,x$.

As is easily demonstrated, the three-dimensional equivalent to Equation (5.43) is

$$\mathbf{f} = -\nabla U \equiv -\left(\frac{\partial U}{\partial x},\,\frac{\partial U}{\partial y},\,\frac{\partial U}{\partial z}\right). \tag{5.44}$$

For example, we have seen that the gravitational potential energy of a mass, m, moving above the Earth's surface is $U = m\,g\,z$, where z measures height above sea level. It follows that the associated gravitational force is

$$\mathbf{f} = (0,\,0,\,-m\,g). \tag{5.45}$$

In other words, the force is of magnitude $m\,g$, and is directed vertically downward. The total energy of the mass shown in Figure 5.8 is the sum of its kinetic and potential energies;

$$E = K + U = K + \frac{1}{2}\,k\,x^2. \tag{5.46}$$

Of course, E remains constant during the mass's motion. Hence, the previous expression can be rearranged to give

$$K = E - \frac{1}{2}\,k\,x^2. \tag{5.47}$$

Because it is impossible for a kinetic energy to be negative, the previous expression suggests that $|x|$ can never exceed the value

$$a = \sqrt{\frac{2\,E}{k}}. \tag{5.48}$$

Here, a is termed the *amplitude* of the mass's motion. Note that when x attains its maximum value a, or its minimum value $-a$, the kinetic energy is momentarily zero (i.e., $K = 0$).

5.7 MOTION IN A GENERAL ONE-DIMENSIONAL POTENTIAL

Suppose that the curve $U(x)$ in Figure 5.9 represents the potential energy of some mass, m, moving in a one-dimensional conservative force-field. For instance, $U(x)$ might represent the gravitational potential energy of a cyclist freewheeling in a hilly region. Note that we have set the potential energy at infinity to zero. This is a useful, and quite common convention (recall that potential energy is undefined to within an arbitrary additive constant). What can we deduce about the motion of the mass in this potential?

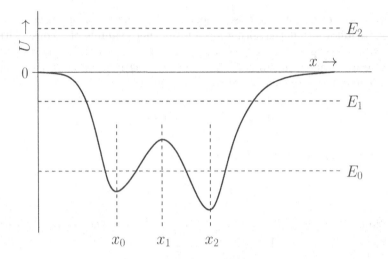

Figure 5.9 General one-dimensional potential. (Reproduced from Fitzpatrick 2012. Courtesy of Cambridge University Press.)

We know that the total energy, E—which is the sum of the kinetic energy, K, and the potential energy, U—is a constant of the motion. Hence, we can write

$$K(x) = E - U(x). \tag{5.49}$$

Now, we also know that a kinetic energy can never be negative, so the previous expression tells us that the motion of the mass is restricted to the region (or regions) in which the potential energy curve $U(x)$ falls below the value E. This idea is illustrated in Figure 5.9. Suppose that the total energy of the system is E_0. It is clear, from the figure, that the mass is trapped inside one or other of the two dips in the potential; these dips are generally referred to as *potential wells*. Suppose that we now raise the energy to E_1. In this case, the mass is free to enter or leave each of the potential wells, but its motion is still bounded to some extent, because it clearly cannot move off to infinity. Finally, let us raise the energy to E_2. Now the mass is unbounded; that is, it can move off to infinity. In systems in which it makes sense to adopt the convention that the potential energy at infinity is zero, bounded systems are characterized by $E < 0$, whereas unbounded systems are characterized by $E > 0$.

The previous discussion suggests that the motion of a mass moving in a potential generally becomes less bounded as the total energy E of the system increases. Conversely, we would expect the motion to become more bounded as E decreases. In fact, if the energy becomes sufficiently small, it appears likely that the system will settle down in some equilibrium state in which the mass is stationary. Let us try to identify any prospective equilibrium states in Figure 5.9. If the mass remains stationary then it must be subject to zero force (otherwise it would accelerate). Hence, according to Equation (5.43), an equilibrium state is characterized by

$$\frac{dU}{dx} = 0. \tag{5.50}$$

In other words, a equilibrium state corresponds to either a maximum or a minimum of the potential energy curve, $U(x)$. It can be seen that the $U(x)$ curve shown in Figure 5.9 has three associated equilibrium states; these are located at $x = x_0$, $x = x_1$, and $x = x_2$.

Let us now make a distinction between *stable* equilibrium points and *unstable* equilibrium points. When the system is slightly perturbed from a stable equilibrium point then the resultant force, f, should always be such as to attempt to return the system to that point. In other words, if $x = x_0$ is an equilibrium point then we require

$$\left.\frac{df}{dx}\right|_{x_0} < 0 \tag{5.51}$$

for stability; that is, if the system is perturbed to the right, so that $x - x_0 > 0$, then the force must act to the left, so that $f < 0$, and vice versa. Likewise, if

$$\left.\frac{df}{dx}\right|_{x_0} > 0 \tag{5.52}$$

then the equilibrium point $x = x_0$ is unstable. It follows, from Equation (5.43), that stable equilibrium points are characterized by

$$\frac{d^2 U}{dx^2} > 0. \tag{5.53}$$

In other words, a stable equilibrium point corresponds to a minimum of the potential energy curve, $U(x)$. Likewise, an unstable equilibrium point corresponds to a maximum of the $U(x)$ curve. Hence, we conclude that $x = x_0$ and $x = x_2$ are stable equilibrium points, in Figure 5.9, whereas $x = x_1$ is an unstable equilibrium point. Of course, this makes perfect sense if we think of $U(x)$ as a gravitational potential energy curve, in which case U is directly proportional to height. All we are saying is that it is easy to confine a low energy mass at the bottom of a valley, but very difficult to balance the same mass on the top of a hill (because any slight perturbation to the mass will cause it to slide down the hill). Note, finally, that if

$$\frac{dU}{dx} = \frac{d^2 U}{dx^2} = 0 \tag{5.54}$$

at any point (or in any region) then we have what is known as a *neutral* equilibrium point. We can move the mass slightly away from such a point, and it will still remain in equilibrium (i.e., it will neither attempt to return to its initial state nor will it continue to move). A neutral equilibrium point corresponds to a flat spot in a $U(x)$ curve. See Figure 5.10.

5.8 POWER

Suppose that an object moves in a general force-field, $\mathbf{f}(\mathbf{r})$. We now know how to calculate how much energy flows from the force-field to the object as it moves along a given path between two points. Let us now consider the rate at which this energy flows. If dW is the amount of work that the force-field performs on the mass in a time interval dt then the rate of working is given by

$$P = \frac{dW}{dt}. \tag{5.55}$$

In other words, the rate of working—which is usually referred to as the *power*—is simply the time derivative of the work performed.

Suppose that the object displaces by $d\mathbf{r}$ in the time interval dt. By definition, the amount of work done on the object during this time interval is given by

$$dW = \mathbf{f} \cdot d\mathbf{r}. \tag{5.56}$$

Stable Equilibrium   Unstable Equilibrium   Neutral Equilibrium

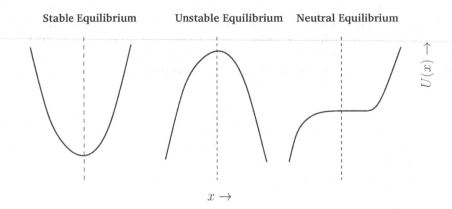

Figure 5.10 Different types of equilibrium. (Reproduced from Fitzpatrick 2012. Courtesy of Cambridge University Press.)

It follows from Equation (5.55) that

$$P = \mathbf{f} \cdot \mathbf{v}, \tag{5.57}$$

where $\mathbf{v} = d\mathbf{r}/dt$ is the object's instantaneous velocity. Note that power can be positive or negative, depending on the relative directions of the vectors $\mathbf{f}$ and $\mathbf{v}$. If these two vectors are mutually perpendicular then the power is zero. For the case of one-dimensional motion, the previous expression reduces to

$$P = f\,v. \tag{5.58}$$

In other words, in one dimension, power simply equals force times velocity.

5.9 EXERCISES

5.1 A person lifts a 30 kg bucket from a well whose depth is 150 m. Assuming that the person lifts the bucket at a constant rate, how much work does he or she perform? [Ans: 4.42×10^4 J.]

5.2 A pirate drags a 50 kg treasure chest over the rough surface of a dock by exerting a constant force of 95 N acting at an angle of 15° above the horizontal. The chest moves 6 m in a straight line, and the coefficient of kinetic friction between the chest and the dock is 0.15.

  (a) How much work does the pirate perform? [Ans: 550.6 J.]

  (b) How much energy is dissipated as heat via friction? [Ans: 419.3 J.]

  (c) What is the final velocity of the chest? [Ans: 2.29 m/s.]

5.3 The force required to slowly stretch a spring varies from 0 N to 105 N as the spring is extended by 13 cm from its unstressed length. Assume that the spring obeys Hooke's law.

  (a) What is the force constant of the spring? [Ans: 807.7 N/m.]

  (b) What work is done in stretching the spring? [Ans: 6.83 J.]

5.4 A roller coaster cart of mass $m = 300\,\text{kg}$ starts at rest at point A, whose height off the ground is $h_1 = 25\,\text{m}$, and a little while later reaches point B, whose height off the ground is $h_2 = 7\,\text{m}$.

  (a) What is the potential energy of the cart relative to the ground at point A? [Ans: $7.36 \times 10^4\,\text{J}$.]

  (b) What is the speed of the cart at point B, neglecting the effect of friction? [Ans: $18.8\,\text{m/s}$.]

5.5 A block of mass $m = 3\,\text{kg}$ starts at rest at a height of $h = 43\,\text{cm}$ on a plane that has an angle of inclination of $\theta = 35°$ with respect to the horizontal. The block slides down the plane, and, upon reaching the bottom, then slides along a horizontal surface. The coefficient of kinetic friction of the block on both surfaces is $\mu = 0.25$. How far does the block slide along the horizontal surface before coming to rest? [Ans: $1.10\,\text{m}$.]

5.6 A car of weight $3000\,\text{N}$ possesses an engine whose maximum power output is $160\,\text{kW}$. The maximum speed of this car on a level road is $35\,\text{m/s}$. Assuming that the resistive force (due to a combination of friction and air resistance) remains constant, what is the car's maximum speed on an incline of 1 in 20 [i.e., an incline whose angle to the horizontal is $\sin^{-1}(1/20)$]? [Ans: $33.90\,\text{m/s}$.]

5.7 If work is done on a particle at a constant rate prove that the velocity acquired in moving a distance x from rest varies as $x^{1/3}$. [From Lamb 1942.]

5.8 (a) Prove that the mean kinetic energy of a particle of mass m moving under a constant force, in any interval of time, is

$$\frac{1}{6}\,m\,(u_1^2 + u_1\,u_2 + u_2^2),$$

  where u_1 and u_2 are the initial and final velocities.

  (b) Show that the mean kinetic energy is greater than that at the middle instance of the interval.

  (c) Show that the mean kinetic energy is less than that of the particle when half-way between its initial and final positions.

[From Lamb 1942.]

CHAPTER 6

Conservation of Momentum

6.1 INTRODUCTION

Up to now, we have analyzed the behavior of dynamical systems that consist of single point masses (i.e., objects whose spatial extent is either negligible, or plays no role in their motion), or arrangements of point masses that are constrained to move together because they are connected via inextensible cables. Let us now broaden our approach in order to take into account systems of point masses that exert forces on one another, but are not necessarily constrained to move together. The classic example of such a multi-component point-mass system is one in which two (or more) freely moving masses collide with one another. The physical concept that plays the central role in the dynamics of multi-component point-mass systems is the **conservation of momentum**.

6.2 TWO-COMPONENT SYSTEMS

The simplest imaginable multi-component dynamical system consists of two point-mass objects that are both constrained to move along the same straight-line. See Figure 6.1. Let x_1 be the displacement of the first object, whose mass is m_1. Likewise, let x_2 be the displacement of the second object, whose mass is m_2. Suppose that the first object exerts a force f_{21} on the second object, whereas the second object exerts a force f_{12} on the first. From Newton's third law of motion, we have

$$f_{12} = -f_{21}. \tag{6.1}$$

Suppose, finally, that the first object is subject to an external force (i.e., a force that originates outside the system) F_1, while the second object is subject to an external force F_2.

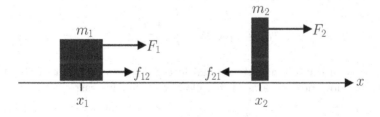

Figure 6.1 A one-dimensional dynamical system consisting of two point mass objects

DOI: 10.1201/9781003198642-6

Applying Newton's second law of motion to each object in turn, we obtain

$$m_1 \ddot{x}_1 = f_{12} + F_1, \tag{6.2}$$

$$m_2 \ddot{x}_2 = f_{21} + F_2. \tag{6.3}$$

Here, ˙ is a convenient shorthand for d/dt. Likewise, ¨ means d^2/dt^2.

At this point, it is helpful to introduce the concept of the *center of mass*. The center of mass is an imaginary point whose displacement, x_{cm}, is defined to be the mass-weighted average of the displacements of the two objects that constitute the system. In other words,

$$x_{cm} = \frac{m_1 x_1 + m_2 x_2}{m_1 + m_2}. \tag{6.4}$$

Thus, if the two object's masses are equal then the center of mass lies halfway between them; if the second object's mass is three times larger than the first then the center of mass lies three-quarters of the way along the straight-line linking the first and second objects, respectively; if the second object's mass is much larger than the first then the center of mass is almost coincident with the second object; and so on.

Summing Equations (6.2) and (6.3), and then making use of Equations (6.1) and (6.4), we obtain

$$m_1 \ddot{x}_1 + m_2 \ddot{x}_2 = (m_1 + m_2) \ddot{x}_{cm} = F_1 + F_2. \tag{6.5}$$

Note that the internal forces, f_{12} and f_{21}, have canceled out. The physical significance of the previous equation becomes clearer if we write it in the following form:

$$M \ddot{x}_{cm} = F, \tag{6.6}$$

where $M = m_1 + m_2$ is the total mass of the system, and $F = F_1 + F_2$ is the net external force acting on the system. Thus, the motion of the center of mass is equivalent to that which would occur if all of the mass contained in the system were collected at the center of mass, and this conglomerate mass were then acted upon by the net external force. In general, this suggests that the motion of the center of mass is simpler than the motions of the component masses, m_1 and m_2. This is particularly the case if the internal forces, f_{12} and f_{21}, are complicated in nature.

Suppose that there are no external forces acting on the system (i.e., $F_1 = F_2 = 0$), or, equivalently, suppose that the sum of all the external forces is zero (i.e., $F = F_1 + F_2 = 0$). In this case, according to Equation (6.6), the motion of the center of mass is governed by Newton's first law of motion; that is, it consists of uniform motion in a straight-line. Hence, in the absence of a net external force, the motion of the center of mass is almost certainly far simpler than that of the component masses.

The velocity of the center of mass is written

$$v_{cm} \equiv \dot{x}_{cm} = \frac{m_1 \dot{x}_1 + m_2 \dot{x}_2}{m_1 + m_2}. \tag{6.7}$$

We have seen that in the absence of external forces v_{cm} is a constant of the motion (i.e., the center of mass does not accelerate). It follows that, in this case,

$$m_1 \dot{x}_1 + m_2 \dot{x}_2 = \text{constant}, \tag{6.8}$$

is also a constant of the motion. Recall, however, from Section 4.3, that **momentum** is defined as the product of mass and velocity. Hence, the momentum of the first mass is

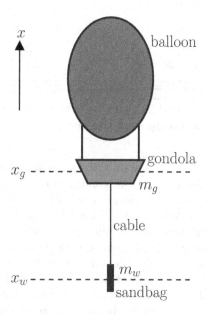

Figure 6.2 An example two-component system

written $p_1 = m_1\,\dot{x}_1$, whereas the momentum of the second mass takes the form $p_2 = m_2\,\dot{x}_2$. It follows that the previous expression corresponds to the total momentum of the system. In other words,

$$P = p_1 + p_2 = \text{constant.} \qquad (6.9)$$

Thus, the total momentum is a conserved quantity; provided there is no net external force acting on the system. This is true irrespective of the nature of the internal forces. More generally, Equation (6.6) can be written

$$\frac{dP}{dt} = F. \qquad (6.10)$$

In other words, the time derivative of the total momentum is equal to the net external force acting on the system; this is just Newton's second law of motion applied to the system as a whole.

6.2.1 Hot-Air Balloon

Let us now apply some of the concepts discussed previously to an example physical system. Consider the simple two-component system shown in Figure 6.2. A gondola of mass m_g hangs from a hot-air balloon whose mass is negligible compared to that of the gondola. A sandbag of mass m_w is suspended from the gondola by means of a light inextensible cable. The system is in equilibrium. Suppose, for the sake of consistency with our other examples, that the x-axis runs vertically upward. Let x_g be the height of the gondola, and x_w, the height of the sandbag. Suppose that the upper end of the cable is attached to a winch inside the gondola, and that this winch is used to slowly shorten the cable, so that the sandbag is lifted upward a distance Δx_w. Does the height of the gondola also change as the cable is reeled in? If so, by how much?

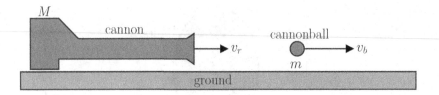

Figure 6.3 Another example two-component system

Let us identify all of the forces acting on the system shown in Figure 6.2. The internal forces are the upward force exerted by the gondola on the sandbag, and the downward force exerted by the sandbag on the gondola. These forces are transmitted via the cable, and are equal and opposite (by Newton's third law of motion). The external forces are the net downward force due to the combined weight of the gondola and the sandbag, and the upward force due to the buoyancy of the balloon. Because the system is in equilibrium, these forces are equal and opposite (it is assumed that the cable is reeled in sufficiently slowly that the equilibrium is not upset). Hence, there is zero net external force acting on the system. It follows, from the previous discussion, that the center of mass of the system is subject to Newton's first law. In particular, because the center of mass is clearly stationary before the winch is turned on, it must remain stationary both during and after the time period during which the winch is operated. Hence, the height of the center of mass,

$$x_{cm} = \frac{m_g\, x_g + m_w\, x_w}{m_g + m_w},\qquad(6.11)$$

is a conserved quantity.

Suppose that the operation of the winch causes the height of the sandbag to change by Δx_w, and that of the gondola to simultaneously change by Δx_g. If x_{cm} is a conserved quantity then we must have

$$0 = m_g\, \Delta x_g + m_w\, \Delta x_w,\qquad(6.12)$$

or

$$\Delta x_g = -\frac{m_w}{m_g}\, \Delta x_w.\qquad(6.13)$$

Thus, if the winch is used to raise the sandbag a distance Δx_w then the gondola is simultaneously pulled downward a distance $(m_w/m_g)\,\Delta x_w$. It is clear that we could use a suspended sandbag as a mechanism for adjusting a hot-air balloon's altitude; the balloon descends as the sandbag is raised, and ascends as it is lowered.

6.2.2 Cannon and Cannonball

Our next example is pictured in Figure 6.3. Suppose that a cannon of mass M propels a cannonball of mass m horizontally with velocity v_b. What is the recoil velocity, v_r, of the cannon? Let us first identify all of the forces acting on the system. The internal forces are the force exerted by the cannon on the cannonball, as the cannon is fired, and the equal and opposite force exerted by the cannonball on the cannon. These forces are extremely large, but only last for a short instance in time; in physics, we term such forces "impulsive". There are no external forces acting in the horizontal direction (which is the only direction that we are considering in this example). It follows that the total (horizontal) momentum, P, of the system is a conserved quantity. Prior to the firing of the cannon, the total momentum is zero (because momentum is mass times velocity, and nothing is initially moving). After

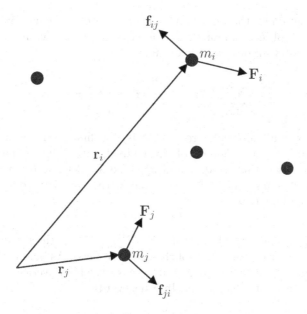

Figure 6.4 A three-dimensional dynamical system consisting of many point-mass objects

the cannon is fired, the total momentum of the system takes the form

$$P = m\,v_b + M\,v_r. \tag{6.14}$$

Because P is a conserved quantity, we can set $P = 0$. Hence,

$$v_r = -\frac{m}{M}\,v_b. \tag{6.15}$$

Thus, the recoil velocity of the cannon is in the opposite direction to the velocity of the cannonball (hence, the minus sign in the previous equation), and is of magnitude $(m/M)\,v_b$. Of course, if the cannon is far more massive that the cannonball (i.e., $M \gg m$), which is usually the case, then the recoil velocity of the cannon is far smaller in magnitude than the velocity of the cannonball. Note, however, that the momentum of the cannon is equal in magnitude to that of the cannonball. It follows that it takes the same effort (i.e., force applied for a certain period of time) to slow down and stop the cannon as it does to slow down and stop the cannonball.

6.3 MULTI-COMPONENT SYSTEMS

Consider a system of N mutually interacting point-mass objects that move in three dimensions. See Figure 6.4. Let the ith object, whose mass is m_i, be located at vector displacement $\mathbf{r}_i$. Suppose that this object exerts a force $\mathbf{f}_{ji}$ on the jth object. By Newton's third law of motion, the force $\mathbf{f}_{ij}$ exerted by the jth object on the ith is given by

$$\mathbf{f}_{ij} = -\mathbf{f}_{ji}. \tag{6.16}$$

Finally, suppose that the ith object is subject to an external force $\mathbf{F}_i$.

Newton's second law of motion applied to the ith object yields

$$m_i\,\ddot{\mathbf{r}}_i = \sum_{\substack{j=1,N \\ j\neq i}} \mathbf{f}_{ij} + \mathbf{F}_i. \tag{6.17}$$

Note that the summation on the right-hand side of the previous equation excludes the case $j = i$, because the ith object cannot exert a force on itself. Let us now take the previous equation and sum it over all objects. We obtain

$$\sum_{i=1,N} m_i \ddot{\mathbf{r}}_i = \sum_{i,j=1,N}^{j \neq i} \mathbf{f}_{ij} + \sum_{i=1,N} \mathbf{F}_i. \tag{6.18}$$

Consider the sum over all internal forces; that is, the first term on the right-hand side of the previous equation. Each element of this sum—$\mathbf{f}_{ij}$, say—can be paired with another element—$\mathbf{f}_{ji}$, in this case—that is equal and opposite. In other words, the elements of the sum all cancel out in pairs. Thus, the net value of the sum is zero. It follows that the previous equation can be written

$$M \ddot{\mathbf{r}}_{cm} = \mathbf{F}, \tag{6.19}$$

where $M = \sum_{i=1,N} m_i$ is the total mass and $\mathbf{F} = \sum_{i=1,N} \mathbf{F}_i$ is the net external force. The quantity $\mathbf{r}_{cm}$ is the vector displacement of the center of mass. As before, the center of mass is an imaginary point whose coordinates are the mass-weighted averages of the coordinates of the objects that constitute the system. In other words,

$$\mathbf{r}_{cm} = \frac{\sum_{i=1,N} m_i \, \mathbf{r}_i}{\sum_{i=1,N} m_i}. \tag{6.20}$$

According to Equation (6.19), the motion of the center of mass is equivalent to that which would be obtained if all of the mass contained in the system were collected at the center of mass, and this conglomerate mass were then acted upon by the net external force. As before, the motion of the center of mass is likely to be far simpler than the motions of the component masses.

Suppose that there is zero net external force acting on the system, so that $\mathbf{F} = \mathbf{0}$. In this case, Equation (6.19) implies that the center of mass moves with uniform velocity in a straight-line. In other words, the velocity of the center of mass,

$$\dot{\mathbf{r}}_{cm} = \frac{\sum_{i=1,N} m_i \, \dot{\mathbf{r}}_i}{\sum_{i=1,N} m_i}, \tag{6.21}$$

is a constant of the motion. The momentum of the ith object takes the form $\mathbf{p}_i = m_i \, \dot{\mathbf{r}}_i$. Hence, the total momentum of the system is written

$$\mathbf{P} = \sum_{i=1,N} m_i \, \dot{\mathbf{r}}_i. \tag{6.22}$$

A comparison of Equations (6.21) and (6.22) suggests that $\mathbf{P}$ is also a constant of the motion when zero net external force acts on the system. Finally, Equation (6.19) can be rewritten

$$\frac{d\mathbf{P}}{dt} = \mathbf{F}. \tag{6.23}$$

In other words, the time derivative of the total momentum is equal to the net external force acting on the system.

It is clear, from the previous discussion, that most of the important results obtained in Section 6.2, for the case of a two-component system moving in one dimension, also apply to a multi-component system moving in three dimensions.

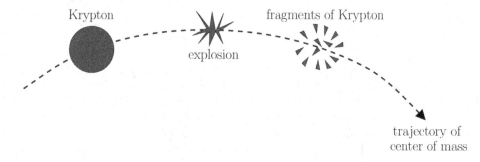

Figure 6.5 The unfortunate history of the planet Krypton

6.3.1 Explosion of Krypton

As an illustration of the points raised in the previous discussion, let us consider the unfortunate history of the planet Krypton. As is well-known, Krypton—Superman's home planet—eventually exploded. Note, however, that before, during, and after this explosion, the net external force acting on Krypton, or the fragments of Krypton—namely, the gravitational attraction due to Krypton's sun—remained the same. In other words, the forces responsible for the explosion can be thought of as large, transitory, internal forces. We conclude that the motion of the center of mass of Krypton, or the fragments of Krypton, was unaffected by the explosion. This follows, from Equation (6.19), because the motion of the center of mass is independent of internal forces. Before the explosion, the planet Krypton presumably executed a standard Keplerian orbit around Krypton's sun. We conclude that, after the explosion, the fragments of Krypton (or, to be more exact, the center of mass of these fragments) continued to execute exactly the same orbit. See Figure 6.5.

6.4 ROCKET SCIENCE

A rocket engine is the only type of propulsion device that operates effectively in outer space. A rocket engine works by ejecting a propellant at high velocity from its rear end. The rocket exerts a backward force on the propellant, in order to eject it, and, by Newton's third law, the propellant exerts an equal and opposite force on the rocket, which moves it forward.

Let us attempt to find the equation of motion of a rocket. Let M be the fixed mass of the rocket engine and the payload, and $m(t)$ the total mass of the propellant contained in the rocket's fuel tanks at time t. Suppose that the rocket engine ejects the propellant at some fixed velocity, u, relative to the rocket. Let us examine the rocket at two closely spaced instances in time. Suppose that at time t the rocket and propellant, whose total mass is $M + m$, are traveling with instantaneous velocity v. Suppose, further, that between times t and $t + dt$ the rocket ejects a quantity of propellant of mass $-dm$ (dm is understood to be negative, so this represents a positive mass) that travels with velocity $v - u$ (i.e., velocity $-u$ in the instantaneous rest frame of the rocket). As a result of the propellant ejection, the velocity of the rocket at time $t + dt$ is boosted to $v + dv$, and its total mass becomes $M + m + dm$. See Figure 6.6.

There is zero external force acting on the system, because the rocket is assumed to be in outer space. It follows that the total momentum of the system is a constant of the motion. Hence, we can equate the momenta evaluated at times t and $t + dt$:

$$(M + m)\,v = (M + m + dm)\,(v + dv) + (-dm)\,(v - u). \tag{6.24}$$

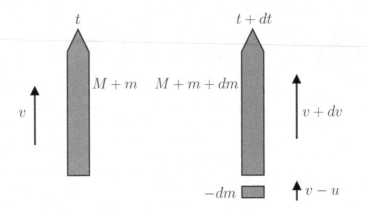

Figure 6.6 Derivation of the rocket equation

Neglecting second-order quantities (i.e., $dm\,dv$), the previous expression yields

$$0 = (M + m)\,dv + u\,dm. \tag{6.25}$$

Rearranging, we obtain

$$\frac{dv}{u} = -\frac{dm}{M + m}. \tag{6.26}$$

Let us integrate the previous equation between an initial time at which the rocket is fully fueled—that is, $m = m_p$, where m_p is the maximum mass of propellant that the rocket can carry—but stationary, and a final time at which the mass of the propellant is m, and the velocity of the rocket is v. Hence,

$$\int_0^v \frac{dv}{u} = -\int_{m_p}^m \frac{dm}{M + m}. \tag{6.27}$$

It follows that

$$\left[\frac{v}{u}\right]_{v=0}^{v=v} = -\left[\ln(M + m)\right]_{m=m_p}^{m=m}, \tag{6.28}$$

which yields

$$v = u \ln\left(\frac{M + m_p}{M + m}\right). \tag{6.29}$$

The final velocity of the rocket (i.e., the velocity attained by the time the rocket has exhausted its fuel, so that $m = 0$) is

$$v_f = u \ln\left(1 + \frac{m_p}{M}\right). \tag{6.30}$$

Note that, unless the initial mass of the propellant exceeds the fixed mass of the rocket by many orders of magnitude (which is highly unlikely), the final velocity, v_f, of the rocket is similar to the velocity, u, with which propellant is ejected from the rear of the rocket in its instantaneous rest frame. This follows because $\ln x$ is similar to unity, unless x becomes extremely large.

Let us now consider the factors that might influence the design of a rocket for use in interplanetary or interstellar travel. Because the distances involved in such travel are vast, it is important that the rocket's final velocity be made as large as possible, otherwise the

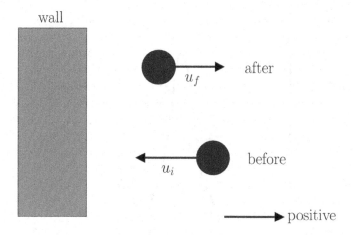

Figure 6.7 A ball bouncing off a wall

journey is going to take an unacceptably long time. However, as we have just seen, the factor that essentially determines the final velocity, v_f, of a rocket is the speed of ejection, u, of the propellant relative to the rocket. Broadly speaking, v_f can never significantly exceed u. It follows that a rocket suitable for interplanetary or interstellar travel should have as high an ejection speed as practically possible. Now, ordinary chemical rockets (the kind that powered the Apollo moon program) can develop enormous thrusts, but are limited to ejection velocities below about $5000\,\text{m/s}$. Such rockets are ideal for lifting payloads out of the Earth's gravitational field, but their relatively low ejection velocities render them unsuitable for long-distance space travel. A new type of rocket engine, called an *ion thruster*, is currently under development; ion thrusters operate by accelerating ions electrostatically to great velocities, and then ejecting them. Although ion thrusters only generate very small thrusts, compared to chemical rockets, their much larger ejection velocities (up to 100 times those of chemical rockets) makes them far more suitable for interplanetary or interstellar space travel. The first spacecraft to employ an ion thruster was the Deep Space 1 probe, which was launched from Cape Canaveral on October 24, 1998; this probe successfully encountered the asteroid 9969 Braille in July, 1999.

6.5   IMPULSES

Suppose that a ball of mass m and speed u_i strikes an immovable wall normally, and rebounds with speed u_f. See Figure 6.7. Clearly, the momentum of the ball is changed by the collision with the wall, because the direction of the ball's velocity is reversed. It follows that the wall must exert a force on the ball, because force is the rate of change of momentum. This force is generally very large, but is only exerted for the short instance in time during which the ball is in physical contact with the wall. As we have already mentioned, physicists generally refer to such a force as an impulsive force.

Figure 6.8 shows the typical time history of an impulsive force, $f(t)$. It can be seen that the force is only non-zero in the short time interval t_1 to t_2. It is helpful to define a quantity known as the net *impulse*, I, associated with $f(t)$:

$$I = \int_{t_1}^{t_2} f(t)\, dt. \tag{6.31}$$

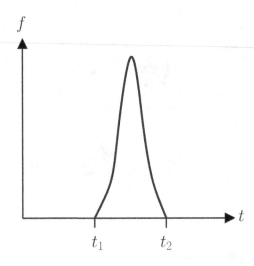

f

t_1 t_2

t

Figure 6.8 An impulsive force

In other words, I is the total area under the $f(t)$ curve shown in Figure 6.8.

Consider a object subject to the impulsive force pictured in Figure 6.8. Newton's second law of motion yields

$$\frac{dp}{dt} = f, \tag{6.32}$$

where p is the momentum of the object. Integrating the previous equation, making use of the definition (6.31), we obtain

$$\Delta p = I. \tag{6.33}$$

Here, $\Delta p = p_f - p_i$, where p_i is the momentum before the impulse, and p_f is the momentum after the impulse. We conclude that the net change in momentum of an object subject to an impulsive force is equal to the total impulse associated with that force. For instance, the net change in momentum of the ball bouncing off the wall in Figure 6.7 is $\Delta p = m\,u_f - m\,(-u_i) = m\,(u_f + u_i)$. (The initial velocity is $-u_i$, because the ball is initially moving in the negative direction.) It follows that the net impulse imparted to the ball by the wall is $I = m\,(u_f + u_i)$. Suppose that we know the ball was only in physical contact with the wall for the short time interval Δt. We conclude that the average force, $\bar{f}$, exerted on the ball during this time interval is

$$\bar{f} = \frac{I}{\Delta t}. \tag{6.34}$$

The previous discussion is only relevant to one-dimensional motion. However, the generalization to three-dimensional motion is fairly straightforward. Consider an impulsive force, $\mathbf{f}(t)$, that is only non-zero in the short time interval t_1 to t_2. The vector impulse associated with this force is simply

$$\mathbf{I} = \int_{t_1}^{t_2} \mathbf{f}(t)\, dt. \tag{6.35}$$

The net change in momentum of an object subject to the force $\mathbf{f}(t)$ is

$$\Delta \mathbf{p} = \mathbf{I}. \tag{6.36}$$

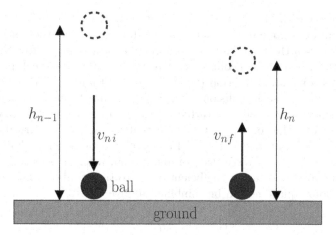

Figure 6.9 A ball bouncing off the ground

Finally, if $t_2 - t_1 = \Delta t$ then the average force experienced by the object in the time interval t_1 to t_2 is

$$\bar{\mathbf{f}} = \frac{\mathbf{I}}{\Delta t}. \tag{6.37}$$

6.6 BOUNCING BALL

Suppose that at time t_0 a ball is thrown vertically upward from ground level with velocity v_{0f}. It is a matter of experience that such a ball will eventually return to the ground, and subsequently bounce off the ground many times before finally coming to rest. Let us try to model this process.

Let v_{ni} be the ball's vertical velocity (upward velocities are positive) immediately before the nth bounce, v_{nf} the ball's vertical velocity immediately after the nth bounce, h_n the maximum height reached by the ball after the nth bounce, and t_n the time of the nth bounce. See Figure 6.9. Making use of the analysis contained in Section 2.8, we find that

$$v_{n+1\,i} = -v_{nf}, \tag{6.38}$$

$$h_n = \frac{v_{nf}^2}{2\,g}, \tag{6.39}$$

$$t_{n+1} = t_n + \frac{2\,v_{nf}}{g}, \tag{6.40}$$

for $n = 0, \infty$.

We now need to relate the ball's vertical velocity immediately before the nth bounce, v_{ni}, to its vertical velocity immediately after the nth bounce, v_{nf}. In order to achieve this goal, Isaac Newton proposed the following empirical law:

$$v_{nf} = -e\,v_{ni}, \tag{6.41}$$

where e is a dimensionless number that typical lies between zero and one. According to this law, which is known as *Newton's experimental law*, and which turns out to be fairly accurate, the ball reverses direction, but loses a fixed fraction, $1 - e^2$, of its kinetic energy during each collision with the ground. Here, e is termed the *coefficient of restitution*. If $e = 1$ then the ball loses none of its kinetic energy during each collision; such collisions are termed *elastic*. If $e = 0$ then the ball loses all of its kinetic energy after the first collision (i.e., it sticks to the ground); such collisions are termed *totally inelastic*. Finally, if $0 < e < 1$ then the ball loses some fixed fraction (which is greater than zero but less than unity) of its kinetic energy during each collision; such collisions are termed *inelastic*. The energy lost in each collision is eventually converted into heat. In general, this conversion is quite a complicated process. Hence, it is remarkable that the simple empirical law (6.41) can accurately capture the conversion process. Incidentally, a table tennis ball bouncing off a steel block has a coefficient of restitution of about 0.90, whereas a leather basketball that bounces off a hardwood floor has a coefficient of restitution of about 0.83.

The previous four equations can be combined to give

$$v_{nf} = e^n \, v_{0f}, \tag{6.42}$$

$$h_n = e^{2n} \, h_0, \tag{6.43}$$

$$t_n = t_0 + \frac{2 \, v_{0f}}{g} \, S_n, \tag{6.44}$$

for $n = 1, \infty$, where

$$S_n = \sum_{n'=1,n} e^{n'-1}. \tag{6.45}$$

However, as is well known, the geometric series S_n can be summed to give

$$S_n = \frac{1 - e^n}{1 - e}. \tag{6.46}$$

Hence,

$$t_n = t_0 + \frac{2 \, v_{0f}}{g} \frac{1 - e^n}{1 - e}. \tag{6.47}$$

According to Equation (6.42), the ball's velocity after the nth bounce is a factor e^n smaller (assuming that $0 \le e < 1$) than the velocity, v_{0f}, with which it was initially thrown upward. Furthermore, according to Equation (6.43), the maximum height attained by the ball after the nth bounce is a factor e^{2n} smaller than the maximum height, h_0, attained before the first bounce. Finally, according to Equation (6.47), the time required for the ball to execute an infinite number of bounces, after which it has lost all of its kinetic energy and come to rest on the ground, is

$$\Delta t = t_\infty - t_0 = \frac{2 \, v_{0f}}{g \, (1 - e)}. \tag{6.48}$$

Note that this time is finite. In other words, after being thrown into the air, and bouncing off the ground many times, the ball eventually comes to rest in a finite time, which is in accordance with our experience. The only exception occurs when $e = 1$, in which case the ball keeps bouncing for ever, because each collision with the ground is elastic, and, therefore, energy conserving.

Let us slightly modify the problem that we have just discussed. Suppose that at $t = t_0$ the ball is thrown from ground level with the vertical velocity component v_{0f}, and the horizontal velocity component u_{0f}. Assuming that there is no friction between the ball and the ground, we would not expect the ball's horizontal motion to affect its vertical motion

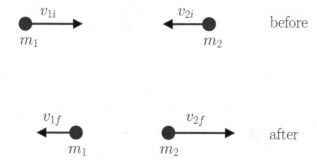

Figure 6.10   A one-dimension collision in the laboratory frame

at all. Hence, the previous analysis remains valid. However, we need to rewrite Newton's experimental law, (6.41), in the form

$$\mathbf{v}_{nf} \cdot \mathbf{n} = -e\,\mathbf{v}_{ni} \cdot \mathbf{n}, \tag{6.49}$$

where $\mathbf{v}_{ni}$ is the ball's velocity vector just before the nth bounce, $\mathbf{v}_{nf}$ is the velocity vector just after the bounce, and $\mathbf{n}$ is a unit vertical vector. In other words, it is only the vertical component of the ball's velocity that reverses, and is reduced in magnitude by a factor e, as a result of the bounce. The ball's horizontal velocity component is unaffected. Here, we can think of $\mathbf{n}$ as the direction of the impulse imparted to the ball when it collides with the ground.

6.7   ONE-DIMENSIONAL COLLISIONS

Consider two objects of mass m_1 and m_2, respectively, that are free to move in one dimension. Suppose that these two objects collide. Suppose, further, that both objects are subject to zero net force when they are not in contact with one another. This situation is illustrated in Figure 6.10.

Both before and after the collision, the two objects move with constant velocity. Let v_{1i} and v_{2i} be the velocities of the first and second objects, respectively, before the collision. Here, velocities to the right in Figure 6.10 are positive. Likewise, let v_{1f} and v_{2f} be the velocities of the first and second objects, respectively, after the collision. During the collision itself, the first object exerts a large transitory force, f_{21}, on the second, whereas the second object exerts an equal and opposite force, $f_{12} = -f_{21}$, on the first. In fact, we can model the collision as equal and opposite impulses given to the two objects at the instant in time when they come together.

We are clearly considering a system in which there is zero net external force (the forces associated with the collision are internal in nature). Hence, the total momentum of the system is a conserved quantity. Equating the total momenta before and after the collision, we obtain

$$m_1\,v_{1i} + m_2\,v_{2i} = m_1\,v_{1f} + m_2\,v_{2f}. \tag{6.50}$$

This equation is valid for any one-dimensional collision, irrespective its nature.

When Newton's experimental law, (6.41), is applied to the case under investigation, it takes the form

$$v_{2f} - v_{1f} = -e\,(v_{2i} - v_{1i}). \tag{6.51}$$

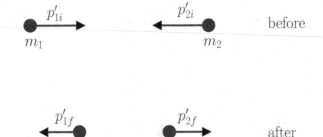

Figure 6.11 A one-dimension collision in the center-of-mass frame

In other words, the relative velocity of the two objects reverses direction during a collision, and is reduced in magnitude by a factor e, where e is the coefficient of restitution for collisions between the objects.

Suppose that we transform to a frame of reference that co-moves with the center of mass of the system. The motion of a multi-component system often looks particularly simple when viewed in such a frame. Because the system is subject to zero net external force, the velocity of the center of mass is invariant, and is given by

$$v_{cm} = \frac{m_1\, v_{1i} + m_2\, v_{2i}}{m_1 + m_2} = \frac{m_1\, v_{1f} + m_2\, v_{2f}}{m_1 + m_2}. \tag{6.52}$$

An object that possesses a velocity v in our original frame of reference—henceforth, termed the *laboratory frame*—possesses a velocity $v' = v - v_{cm}$ in the so-called *center-of-mass frame*. It is easily demonstrated that

$$v'_{1i} = -\frac{m_2}{m_1 + m_2}\,(v_{2i} - v_{1i}), \tag{6.53}$$

$$v'_{2i} = +\frac{m_1}{m_1 + m_2}\,(v_{2i} - v_{1i}), \tag{6.54}$$

$$v'_{1f} = -\frac{m_2}{m_1 + m_2}\,(v_{2f} - v_{1f}), \tag{6.55}$$

$$v'_{2f} = +\frac{m_1}{m_1 + m_2}\,(v_{2f} - v_{1f}). \tag{6.56}$$

The previous four equations yield

$$-p'_{1i} = p'_{2i} = \mu\,(v_{2i} - v_{1i}), \tag{6.57}$$

$$-p'_{1f} = p'_{2f} = \mu\,(v_{2f} - v_{1f}), \tag{6.58}$$

where $\mu = m_1\, m_2/(m_1 + m_2)$ is the so-called *reduced mass* (see Section 14.12), and $p'_{1f} = m_1\, v'_{1i}$ is the initial momentum of the first object in the center-of-mass frame, etcetera. In other words, when viewed in the center-of-mass frame, the two objects approach one another with equal and opposite momentum before the collision, and diverge from one another with equal and opposite momentum after the collision. See Figure 6.11. Thus, the center-of-mass momentum conservation equation,

$$p'_{1i} + p'_{2i} = p'_{1f} + p'_{2f}, \tag{6.59}$$

is trivially satisfied, because both the left- and right-hand sides are zero. Incidentally, this result is valid for both elastic and inelastic collisions.

Equations (6.51), (6.57), and (6.58) can be combined to give

$$p'_{1f} = -e\,p'_{1i},\tag{6.60}$$

$$p'_{2f} = -e\,p'_{2i}.\tag{6.61}$$

In other words, in the center-of-mass frame, the collision causes the equal and opposite momenta of the two objects to reverse direction, and reduce in magnitude by a factor e. The previous two expressions imply that

$$v'_{1f} = -e\,v'_{1i},\tag{6.62}$$

$$v'_{2f} = -e\,v'_{2i}.\tag{6.63}$$

In other words, in the center-of-mass frame, the collision also causes the velocities of the two objects to reverse direction, and reduce in magnitude by a factor e. It follows that the kinetic energy of the system in the center-of-mass frame is reduced by a factor e^2 as a result of the collision.

Equations (6.53) and (6.54) can be combined with the previous two equations to give

$$v'_{1f} = \frac{e\,m_2}{m_1 + m_2}\,(v_{2i} - v_{1i}),\tag{6.64}$$

$$v'_{2f} = -\frac{e\,m_1}{m_1 + m_2}\,(v_{2i} - v_{1i}).\tag{6.65}$$

However, $v_{1f} = v'_{1f} + v_{cm}$ and $v_{2f} = v'_{2f} + v_{cm}$, which allows us to express the velocities of the two objects after the collision in the laboratory frame in terms of the corresponding velocities before the collision:

$$v_{1f} = \left(\frac{m_1 - e\,m_2}{m_1 + m_2}\right) v_{1i} + \left[\frac{(1+e)\,m_2}{m_1 + m_2}\right] v_{2i},\tag{6.66}$$

$$v_{2f} = \left[\frac{(1+e)\,m_1}{m_1 + m_2}\right] v_{1i} - \left(\frac{e\,m_1 - m_2}{m_1 + m_2}\right) v_{2i}.\tag{6.67}$$

Given that

$$p_1 = p'_1 + m_1\,v_{cm},\tag{6.68}$$

$$p_2 = p'_2 + m_2\,v_{cm},\tag{6.69}$$

the kinetic energy of the system in the laboratory frame can be written

$$K \equiv \frac{p_1^2}{2\,m_1} + \frac{p_2^2}{2\,m_2} = \frac{(p'_1 + m_1\,v_{cm})^2}{2\,m_1} + \frac{(p'_2 + m_2\,v_{cm})^2}{2\,m_2}$$

$$= K' + \frac{1}{2}\,(m_1 + m_2)\,v_{cm}^2,\tag{6.70}$$

where

$$K' = \frac{(p'_1)^2}{2\,m_1} + \frac{(p'_2)^2}{2\,m_2}\tag{6.71}$$

is the kinetic energy in the center-of-mass frame, and use has been made of the fact that $p'_1 = -p'_2$. However, we know that

$$K'_f = e^2\,K'_i,\tag{6.72}$$

where the subscripts i and f refer to the kinetic energies before and after the collision, respectively. It follows that

$$K_i = K_i' + \frac{1}{2}(m_1 + m_2) v_{cm}^2, \tag{6.73}$$

$$K_f = e^2 K_i' + \frac{1}{2}(m_1 + m_2) v_{cm}^2, \tag{6.74}$$

which implies that

$$K_f = e^2 K_i + \frac{1}{2}(1 - e^2)(m_1 + m_2) v_{cm}^2. \tag{6.75}$$

Thus, in the laboratory frame, the fraction of the initial kinetic energy lost in the collision is

$$f \equiv \frac{K_i - K_f}{K_i} = (1 - e^2) \left[1 - \frac{(m_1 + m_2) v_{cm}^2}{2 K_i} \right]$$

$$= (1 - e^2) \frac{m_1 m_2}{m_1 + m_2} \frac{(v_{2i} - v_{1i})^2}{(m_1 v_{1i}^2 + m_2 v_{2i}^2)}. \tag{6.76}$$

6.7.1 Elastic Collisions

Consider an elastic collision. In other words, suppose that the coefficient of restitution takes the value $e = 1$. Equations (6.66) and (6.67) reduce to:

$$v_{1f} = \left(\frac{m_1 - m_2}{m_1 + m_2} \right) v_{1i} + \left(\frac{2 m_2}{m_1 + m_2} \right) v_{2i}, \tag{6.77}$$

$$v_{2f} = \left(\frac{2 m_1}{m_1 + m_2} \right) v_{1i} - \left(\frac{m_1 - m_2}{m_1 + m_2} \right) v_{2i}. \tag{6.78}$$

Moreover, Equation (6.76) yields $f = 0$. In other words, there is no energy loss in an elastic collision.

Let us, now, consider some special cases. Suppose that two equal-mass objects collide elastically. If $m_1 = m_2$ then Equations (6.77) and (6.78) yield

$$v_{1f} = v_{2i}, \tag{6.79}$$

$$v_{2f} = v_{1i}. \tag{6.80}$$

In other words, the two objects simply exchange velocities when they collide. For instance, if the second object is stationary and the first object strikes it head-on with velocity v then the first object is brought to a halt whereas the second object moves off with velocity v. It is possible to reproduce this effect in pool by striking the cue ball with great force in such a manner that it slides, rather that rolls, over the table; in this case, when the cue ball strikes another ball head-on it comes to a complete halt, and the other ball is propelled forward very rapidly. Incidentally, it is necessary to prevent the cue ball from rolling, because rolling motion is not taken into account in our analysis, and actually changes the answer.

Suppose that the second object is much more massive than the first (i.e., $m_2 \gg m_1$), and is initially at rest (i.e., $v_{2i} = 0$). In this case, Equations (6.77) and (6.78) yield

$$v_{1f} \simeq -v_{1i}, \tag{6.81}$$

$$v_{2f} \simeq 0. \tag{6.82}$$

In other words, the velocity of the light object is effectively reversed during the collision, whereas the massive object remains approximately at rest. Indeed, this is the sort of behavior we expect when an object collides elastically with an immovable obstacle; for instance, when an elastic ball bounces off a brick wall.

Suppose, finally, that the second object is much lighter than the first (i.e., $m_2 \ll m_1$), and is initially at rest (i.e., $v_{2i} = 0$). In this case, Equations (6.77) and (6.78) yield

$$v_{1f} \simeq v_{1i}, \tag{6.83}$$

$$v_{2f} \simeq 2 v_{1i}. \tag{6.84}$$

In other words, the motion of the massive object is essentially unaffected by the collision, whereas the light object ends up moving twice as fast as the massive one.

6.7.2 Totally Inelastic Collisions

In a totally inelastic collision (i.e., $e = 0$), the two objects stick together after colliding, so they end up moving with the same final velocity, $v_f = v_{1f} = v_{2f}$. In this case, Equations (6.66) and (6.67) yield

$$v_f = \frac{m_1 v_{1i} + m_2 v_{2i}}{m_1 + m_2} = v_{cm}. \tag{6.85}$$

In other words, the common final velocity of the two objects is equal to the centre-of-mass velocity of the system. This is hardly a surprising result. We have already seen that in the centre-of-mass frame the two objects must diverge with equal and opposite momentum after the collision. However, in a totally inelastic collision these two momenta must also be equal (because the two objects stick together). The only way in which this is possible is if the two objects remain stationary in the centre-of-mass frame after the collision. Hence, the two objects move with the centre-of-mass velocity in the laboratory frame.

Suppose that the second object is initially at rest (i.e., $v_{2i} = 0$). In this special case, the common final velocity of the two objects is

$$v_f = \frac{m_1}{m_1 + m_2} v_{1i}. \tag{6.86}$$

Note that the first object is slowed down by the collision. According to Equation (6.76), the fractional loss in kinetic energy of the system due to the collision is given by

$$f = \frac{m_2}{m_1 + m_2}. \tag{6.87}$$

The loss in kinetic energy is small if the (initially) stationary object is much lighter than the moving object (i.e., if $m_2 \ll m_1$), and almost 100% if the moving object is much lighter than the stationary one (i.e., if $m_2 \gg m_1$). Of course, the lost kinetic energy of the system is converted into some other form of energy; for instance, heat energy.

6.7.3 Inelastic Collisions

Consider an inelastic collision for which $0 < e < 1$.

Suppose that both colliding objects have the same mass. If $m_1 = m_2$ then Equations (6.66) and (6.67) yield

$$v_{1f} = \left(\frac{1-e}{2}\right) v_{1i} + \left(\frac{1+e}{2}\right) v_{2i}, \tag{6.88}$$

$$v_{2f} = \left(\frac{1+e}{2}\right) v_{1i} + \left(\frac{1-e}{2}\right) v_{2i}. \tag{6.89}$$

In particular, if the second mass is initially at rest then

$$v_{1f} = \left(\frac{1-e}{2}\right) v_{1i}, \tag{6.90}$$

$$v_{2f} = \left(\frac{1+e}{2}\right) v_{1i}. \tag{6.91}$$

It can be seen that the collision causes the initially moving object to slow down, and imparts motion to the initially stationary object. Moreover, according to Equation (6.76), the factional energy loss in the collision is

$$f = \frac{1-e^2}{2}. \tag{6.92}$$

Suppose that the second object is much more massive than the first (i.e., $m_2 \gg m_1$), and is initially at rest (i.e., $v_{2i} = 0$). In this case, Equations (6.66) and (6.67) yield

$$v_{1f} \simeq -e\, v_{1i}, \tag{6.93}$$

$$v_{2f} \simeq 0. \tag{6.94}$$

This state of affairs is analogous to that discussed in Section 6.6 in which an object collides inelastically with an unyielding surface.

Suppose, finally, that the second object is much lighter than the first (i.e., $m_2 \ll m_1$), and is initially at rest (i.e., $v_{2i} = 0$). In this case, Equations (6.66) and (6.67) yield

$$v_{1f} \simeq v_{1i}, \tag{6.95}$$

$$v_{2f} \simeq (1+e)\, v_{1i}. \tag{6.96}$$

In other words, the motion of the massive object is essentially unaffected by the collision, whereas the light object ends up moving faster than the massive one.

6.8   TWO-DIMENSIONAL COLLISIONS

Suppose that an object of mass m_1, moving with initial velocity $\mathbf{v}_{1i}$, strikes a second object, of mass m_2, that is initially at rest. Suppose, further, that the collision is not head-on, so that after the collision the first object moves off at an angle θ_1 to its initial direction of motion, whereas the second object recoils at an angle θ_2 to this direction. Let the final velocities of the two objects be $\mathbf{v}_{1f}$ and $\mathbf{v}_{2f}$, respectively. See Figure 6.12.

We are again considering a system in which there is zero net external force (the forces associated with the collision are internal in nature). It follows that the total momentum of the system is a conserved quantity. However, unlike before, we must now treat momentum as a vector quantity, because we are no longer dealing with one-dimensional motion. Momentum conservation implies that

$$m_2\, \mathbf{v}_{1i} = m_1\, \mathbf{v}_{1f} + m_2\, \mathbf{v}_{2f}. \tag{6.97}$$

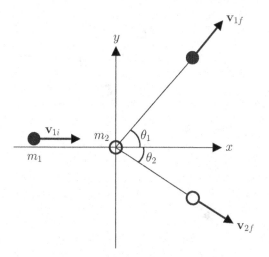

Figure 6.12 A two-dimensional collision in the laboratory frame

As before, it is convenient to transform to a frame of reference that co-moves with the center of mass of the system. The invariant velocity of the center of mass is given by

$$\mathbf{v}_{cm} = \frac{m_1\,\mathbf{v}_{1i}}{m_1 + m_2} = \frac{m_1\,\mathbf{v}_{1f} + m_2\,\mathbf{v}_{2f}}{m_1 + m_2}. \tag{6.98}$$

An object that possesses a velocity $\mathbf{v}$ in the laboratory frame possesses a velocity $\mathbf{v}' = \mathbf{v} - \mathbf{v}_{cm}$ in the center-of-mass frame. Hence, it follows that

$$\mathbf{v}'_{1i} = \left(\frac{m_2}{m_1 + m_2}\right)\mathbf{v}_{1i}, \tag{6.99}$$

$$\mathbf{v}'_{2i} = -\left(\frac{m_1}{m_1 + m_2}\right)\mathbf{v}_{1i}, \tag{6.100}$$

$$\mathbf{v}'_{1f} = -\left(\frac{m_2}{m_1 + m_2}\right)(\mathbf{v}_{2f} - \mathbf{v}_{1f}), \tag{6.101}$$

$$\mathbf{v}'_{2f} = \left(\frac{m_1}{m_1 + m_2}\right)(\mathbf{v}_{2f} - \mathbf{v}_{1f}). \tag{6.102}$$

Furthermore, the momenta in the center-of-mass frame take the form

$$-\mathbf{p}'_{1i} = \mathbf{p}'_{2i} = -\mu\,\mathbf{v}_{1i}, \tag{6.103}$$

$$-\mathbf{p}'_{1f} = \mathbf{p}'_{2f} = -\mu\,(\mathbf{v}_{2f} - \mathbf{v}_{1f}), \tag{6.104}$$

where $\mu = m_1\,m_2/(m_1 + m_2)$. (Of course, $\mathbf{p}'_{1i} = m_1\,\mathbf{v}'_{1i}$, et cetera.) As before, in the center-of-mass frame, the two objects approach one another with equal and opposite momentum before the collision, and diverge from one another with equal and opposite momenta after the collision. Let θ be the direction subtended between the final and initial momenta of each object in the center-of-mass frame. See Figure 6.13. It follows that in the x-y coordinate

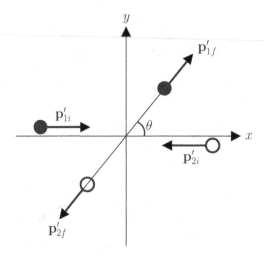

Figure 6.13   A two-dimensional collision in the center-of-mass frame

system shown in the figure,

$$\mathbf{p}'_{1i} = p'_{1i} (1, 0), \tag{6.105}$$

$$\mathbf{p}'_{2i} = -p'_{1i} (1, 0), \tag{6.106}$$

$$\mathbf{p}'_{1f} = p'_{1f} (\cos\theta, \sin\theta), \tag{6.107}$$

$$\mathbf{p}'_{2f} = -p'_{1f} (\cos\theta, \sin\theta), \tag{6.108}$$

where $p'_{1i} = |\mathbf{p}'_{1i}|$, etcetera.

The impulse imparted to the first object as a result of the collision is

$$\mathbf{I}_1 \equiv \mathbf{p}'_{1f} - \mathbf{p}'_{1i} = (p'_{1f} \cos\theta - p'_{1i}, \, p'_{1f} \sin\theta). \tag{6.109}$$

An equal and opposite impulse is imparted to the second object. According to the discussion at the end of Section 6.6, Newton's experimental law implies that

$$\mathbf{p}'_{1f} \cdot \mathbf{I}_1 = -e\,\mathbf{p}'_{1i} \cdot \mathbf{I}_1, \tag{6.110}$$

where e is the coefficient of restitution. In other words, in the center-of-mass frame, the collision reverses the component of the first object's momentum parallel to the direction of the collisional impulse, and reduces the magnitude of the component by a factor e. The same is true of the second object's momentum. Of course, the components of the first and second object's momentum perpendicular to the direction of the impulse are unchanged by the collision. Let

$$g = \frac{p'_{1f}}{p'_{1i}}. \tag{6.111}$$

It follows that the kinetic energy of the system in the center-of-mass frame is reduced by a factor g^2 by the collision. In this respect, g can be thought of as the effective coefficient of restitution for the collision. Equations (6.105), (6.107), and (6.109)–(6.111), can be combined to give

$$g^2 - (1 - e) \cos\theta\, g - e = 0, \tag{6.112}$$

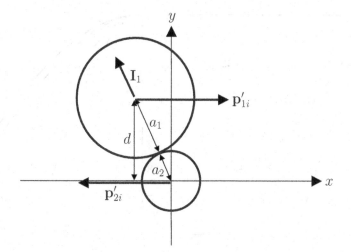

Figure 6.14 Geometry of a two-dimensional collision in the center-of-mass frame

which implies that

$$g = \frac{(1-e)\cos\theta + \sqrt{(1-e)^2\cos^2\theta + 4e}}{2}. \tag{6.113}$$

Note that $g = 1$ for a glancing collision, for which $\theta = 0$, and reduces monotonically with increasing θ (assuming that $e < 1$) until it attains the value e for a head-on collision, for which $\theta = \pi$.

Suppose that the first object is a smooth, hard sphere of radius a_1, whereas the second object is a smooth, hard sphere of radius a_2. Suppose, further, that the unperturbed (by the collision) parallel trajectories of the two objects in the center-of-mass frame (and also in the laboratory frame) are located at a perpendicular distance d apart. As is clear from Figure 6.14, if $d < a_1 + a_2$ then the two objects will collide, otherwise they will miss one another completely. Let us define the *impact parameter*,

$$\mu = \frac{d}{a_1 + a_2}. \tag{6.114}$$

Suppose that $0 \leq \mu < 1$, which ensures that the two objects collide. If the two objects are smooth then the impulse $\mathbf{I}_1$ is parallel to the line joining their centers when they first touch one another, as shown in Figure 6.14. (If the objects are rough then, in general, there will be a frictional component of the impulse that is perpendicular to the line joining the centers of the objects.) Note that the impulse imparts zero torque to the two spheres. Hence, we do not have take rotational motion into account in our analysis (assuming that the two spheres have zero initial rotation). (This would not be true if the objects were rough.) It follows from the figure that

$$\frac{I_{1y}}{I_1} = \frac{d}{a_1 + a_2} = \mu, \tag{6.115}$$

or

$$\mu = \frac{g\sin\theta}{\sqrt{1 - 2g\cos\theta + g^2}}, \tag{6.116}$$

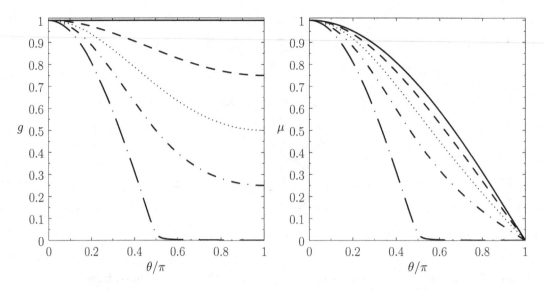

Figure 6.15 The effective coefficient of restitution, g, and the impact parameter, μ, as functions of the center-of-mass scattering angle, θ. The solid, dashed, dotted, short-dash-dotted, and long-dash-dotted curves correspond to $e = 1$, 0.75, 0.5, 0.25, and 0.001, respectively.

where use has been made of Equations (6.109) and (6.111). The previous equation implicitly determines the scattering angle in the center-of-mass frame, θ, in terms of the impact parameter, μ.

Figure 6.15 shows the effective coefficient of restitution, g, and the impact parameter, μ, as functions of the center-of-mass scattering angle, θ, for various values of the true impact parameter, e. It can be seen that the effective coefficient of restitution is a monotonically decreasing function of the scattering angle. This implies that there is very little energy loss associated with a glancing collision (i.e., $\theta \ll 1$), and a significant energy loss associated with a collision that is more head-on in nature (i.e., $\theta \simeq \pi$). Of course, g is unity for elastic collisions (i.e., $e = 1$), and decreases monotonically with decreasing e. It can also be seen that the impact parameter is a monotonically decreasing function of the scattering angle. This implies that large scattering angles correspond to small impact parameters, and vice versa. For elastic collisions (with $g = e = 1$), $\mu = \cos(\theta/2)$. Note that a totally inelastic collision (i.e., $g = 0$) is only possible when the impact parameter is zero, which corresponds to a head-on collision.

In the x-y coordinate system shown in Figure 6.13,

$$\mathbf{p}'_{1f} = g\,p'_{1i}\,(\cos\theta,\,\sin\theta), \tag{6.117}$$

$$\mathbf{p}'_{2f} = -g\,p'_{1i}\,(\cos\theta,\,\sin\theta). \tag{6.118}$$

Hence,

$$\mathbf{v}'_{1f} \equiv \frac{\mathbf{p}'_{1f}}{m_1} = \left(\frac{g\,m_2\,v_{1i}}{m_1 + m_2}\right)(\cos\theta,\,\sin\theta), \tag{6.119}$$

$$\mathbf{v}'_{2f} \equiv \frac{\mathbf{p}'_{2f}}{m_2} = -\left(\frac{g\,m_1\,v_{1i}}{m_1 + m_2}\right)(\cos\theta,\,\sin\theta), \tag{6.120}$$

where use has been made of Equation (6.103).

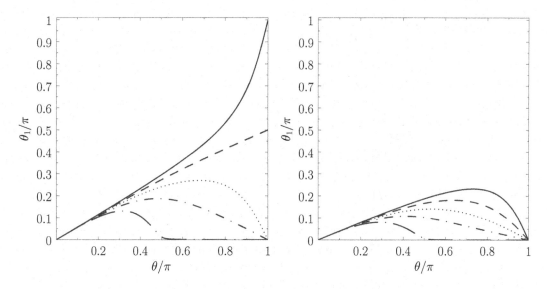

Figure 6.16 The laboratory-frame scattering angle, θ_1, of the initially moving object as a function of the center-of-mass scattering angle, θ. The solid, dashed, dotted, short-dash-dotted, and long-dash-dotted curves correspond to $e = 1$, 0.75, 0.5, 0.25, and 0.001, respectively. The left panel shows the case $m_1/m_2 = 0.75$. The right panel shows the case $m_1/m_2 = 1.5$.

Now, $\mathbf{v} = \mathbf{v}' + \mathbf{v}_{cm}$, where

$$\mathbf{v}_{cm} = \left(\frac{m_1 v_{1i}}{m_1 + m_2}\right)(1, 0). \tag{6.121}$$

It follows that, in the x-y coordinate system shown in Figure 6.12, the laboratory-frame velocities of the two objects after the collision are

$$\mathbf{v}_{1f} = \left(\frac{v_{1i}}{m_1 + m_2}\right)(m_1 + g\, m_2 \cos\theta,\, g\, m_2 \sin\theta), \tag{6.122}$$

$$\mathbf{v}_{2f} = \left(\frac{m_1 v_{1i}}{m_1 + m_2}\right)(1 - g \cos\theta,\, -g \sin\theta). \tag{6.123}$$

Hence, according to Figure 6.12,

$$\tan\theta_1 = \frac{g \sin\theta}{g \cos\theta + m_1/m_2}, \tag{6.124}$$

$$\tan\theta_2 = \frac{g \sin\theta}{1 - g \cos\theta}. \tag{6.125}$$

Figure 6.16 shows the laboratory-frame scattering angle, θ_1, of the initially moving object as a function of the center-of-mass scattering angle, θ, the true coefficient of restitution, e, and the mass ratio, m_1/m_2. It can be seen that backward scattering (i.e., $\theta_1 > \pi/2$) is only possible when the initially moving object is lighter than the initially stationary object (i.e., $m_1 < m_2$), and the true coefficient of restitution, e, is not too small. The scattering angle decreases monotonically as the true coefficient of restitution increases.

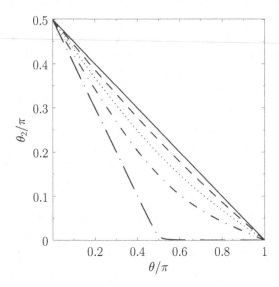

Figure 6.17 The laboratory-frame recoil angle, θ_2, of the initially stationary object as a function of the center-of-mass scattering angle, θ. The solid, dashed, dotted, short-dash-dotted, and long-dash-dotted curves correspond to $e = 1$, 0.75, 0.5, 0.25, and 0.001, respectively.

Figure 6.17 shows the laboratory-frame recoil angle, θ_2, of the initially stationary object as a function of the center-of-mass scattering angle, θ, and the true coefficient of restitution, e. It can be seen that backward recoil (i.e., $\theta_2 > \pi/2$) is impossible. The recoil angle takes its maximum value $\pi/2$ when the collision is glancing (i.e., $\theta = 0$), and decreases monotonically with increasing θ, until it attains the value zero for a head-on collision (i.e., $\theta = \pi$). For elastic collisions, with $e = 1$, we have $\theta_2 = (\pi - \theta)/2$. The recoil angle also decreases monotonically as the true coefficient of restitution increases.

Let

$$K_i = \frac{1}{2}\, m_1\, v_{1i}^2 \tag{6.126}$$

be the kinetic energy of the system in the laboratory frame before the collision. Likewise, let

$$K_{1f} = \frac{1}{2}\, m_1\, v_{1f}^2, \tag{6.127}$$

$$K_{2f} = \frac{1}{2}\, m_1\, v_{2f}^2 \tag{6.128}$$

be the corresponding kinetic energies of the initially moving and initially stationary objects, respectively, after the collision. It follows from Equations (6.122) and (6.123) that

$$K_{1f} = \left[\frac{(m_1/m_2)^2 + 2\,g\,(m_1/m_2)\,\cos\theta + g^2}{(1 + m_1/m_2)^2} \right] K_i, \tag{6.129}$$

$$K_{2f} = \left[\frac{(m_1/m_2)\,(1 - 2\,g\,\cos\theta + g^2)}{(1 + m_1/m_2)^2} \right] K_i. \tag{6.130}$$

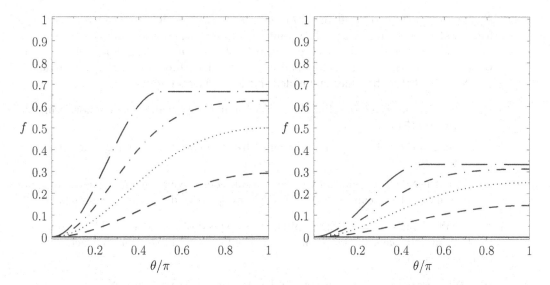

Figure 6.18 The laboratory-frame fraction of kinetic energy lost in the collision as a function of the center-of-mass scattering angle, θ. The solid, dashed, dotted, short-dash-dotted, and long-dash-dotted curves correspond to $e = 1$, 0.75, 0.5, 0.25, and 0.001, respectively. The left and right panels show cases for which $m_1/m_2 = 0.5$ and 2, respectively.

Thus, the total kinetic energy in the laboratory frame after the collision is

$$K_f \equiv K_{1f} + K_{2f} = \left[\frac{(m_1/m_2)^2 + (1 + g^2)(m_1/m_2) + g^2}{(1 + m_1/m_2)^2} \right] K_i. \qquad (6.131)$$

It follows that, in the laboratory frame, the fraction of the initial kinetic energy lost in the collision is

$$f \equiv \frac{K_i - K_f}{K_i} = \frac{1 - g^2}{1 + m_1/m_2}. \qquad (6.132)$$

Figure 6.18 shows the fraction of kinetic energy lost in the collision as a function of the center-of-mass scattering angle, θ, the true coefficient of restitution, e, and the mass ratio, m_1/m_2. It can be seen that the energy loss is greater when the initially moving object is lighter than the initially stationary object. The energy loss is very small for glancing collisions (i.e., $\theta \ll 1$), and increases monotonically with increasing θ, attaining its maximum value for head-on collisions (i.e., $\theta = \pi$). The energy loss is zero for elastic collisions (i.e., $e = 0$), and increases monotonically as the true coefficient of restitution, e, increases.

The analysis in this section reveals that we can fully characterize a collision between two hard-sphere objects, one of which is initially stationary, given the ratio of the masses of the two objects, the velocity of the initially moving object, the coefficient of restitution for the collision, and the impact parameter (which determines the center-of-mass scattering angle). The analysis can easily be generalized to deal with more complicated collisions.

6.9 EXERCISES

6.1 A gun of mass M discharges a shot of mass m horizontally. The energy of the explosion is such as would be sufficient to project the shot vertically to a height h. Find the recoil speed of the gun. [Ans: $\sqrt{2\,g\,h\,m^2/[M\,(M+m)]}$.] [From Lamb 1942.]

6.2 A cannon is bolted to the floor of a railway carriage, which is free to move without friction along a straight track. The combined mass of the cannon and the carriage is $M = 1200\,\text{kg}$. The cannon fires a cannonball, of mass $m = 1.2\,\text{kg}$, horizontally with velocity $v = 115\,\text{m/s}$. The cannonball travels the length of the carriage, a distance $L = 85\,\text{m}$, and then becomes embedded in the carriage's end wall.

(a) What is the recoil speed of the carriage immediately after the cannon is fired? [Ans: 0.115 m/s.]

(b) What is the velocity of the carriage after the cannonball strikes the far wall? [Ans: 0 m/s.]

(c) What net distance, and in what direction, does the carriage move as a result of the firing of the cannon? [Ans: 8.49 cm in opposite direction to cannonball.]

6.3 A softball of mass $m = 0.35\,\text{kg}$ is pitched at a speed of $u = 12\,\text{m/s}$. The hitter hits the ball directly back to the pitcher at a speed of $v = 21\,\text{m/s}$. The bat acts on the ball for $t = 0.01\,\text{s}$.

(a) What impulse is imparted by the bat to the ball? [Ans: 11.55 N s.]

(b) What average force is exerted by the bat on the ball? [Ans: 1155.0 N.]

6.4 A skater of mass $M = 120\,\text{kg}$ is skating across a pond with uniform velocity $v = 8\,\text{m/s}$. One of the skater's friends, who is standing at the edge of the pond, throws a medicine ball of mass $m = 20\,\text{kg}$ with velocity $u = 3\,\text{m/s}$ to the skater, who catches it. The direction of motion of the ball is perpendicular to the initial direction of motion of the skater. Assume that the skater moves without friction.

(a) What is the final speed of the skater? [Ans: 6.87 m/s.]

(b) What is the final direction of motion of the skater relative to his/her initial direction of motion? [Ans: 3.58°.]

6.5 Consider a rocket launched vertically upward from the surface of the Earth. Let the Earth's gravitational acceleration take the uniform value g. Suppose that the rocket ejects mass at the uniform rate α at a speed u in the rocket's instantaneous rest frame. Let m_i be the initial mass of the rocket, and let m_f be the final mass, immediately after the rocket has exhausted its fuel.

(a) Show that the rocket's final velocity, immediately after it has exhausted its fuel, is

$$v_f = -\frac{(m_i - m_f)\,g}{\alpha} + u\ln\left(\frac{m_i}{m_f}\right).$$

(b) Show that the rocket's height off the ground immediately after it has exhausted its fuel is

$$h_f = -\frac{(m_i - m_f)^2\,g}{2\,\alpha^2} + \frac{u}{\alpha}\left[m_f\left(\frac{m_f}{m_i}\right) + m_i - m_f\right].$$

(c) What is is minimum mass ejection rate needed for the rocket to lift off from the ground? [Ans: $m_i\,g/u$.]

6.6 A bullet of mass $m = 12\,\text{g}$ strikes a stationary wooden block of mass $M = 5.2\,\text{kg}$ standing on a frictionless surface. The block, with the bullet embedded in it, acquires a velocity of $v = 1.7\,\text{m/s}$.

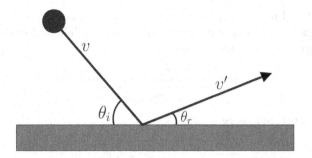

Figure 6.19 Figure for Exercise 6.9

(a) What was the velocity of the bullet before it struck the block? [Ans: 738.4 m/s.]

(b) What fraction of the bullet's initial kinetic energy is lost (i.e., dissipated) due to the collision with the block? [Ans: 0.998.]

6.7 An object of mass $m_1 = 2$ kg, moving with velocity $v_{1i} = 12$ m/s, collides head-on with a stationary object whose mass is $m_2 = 6$ kg. Given that the collision is elastic, what are the final velocities of the two objects. Neglect friction. [Ans: -6 m/s; $+6$ m/s.]

6.8 Two objects slide over a frictionless horizontal surface. The first object, mass $m_1 = 5$ kg, is propelled with speed $v_{1i} = 4.5$ m/s toward the second object, mass $m_2 = 2.5$ kg, which is initially at rest. After the collision, both objects have velocities that are directed $\theta = 30°$ on either side of the original line of motion of the first object.

(a) What are the final speeds of the two objects? [Ans: 2.598 m/s; 5.196 m/s.]

(b) Is the collision elastic or inelastic? [Ans: Elastic.]

6.9 A billiard ball bounces off the cushion of a billiard table. Let v and θ_i be the initial speed and angle of incidence of the ball, and let v' and θ_r be the final speed and angle of reflection of the ball. See Figure 6.19. Let e be the coefficient of restitution between the ball and the cushion.

(a) Derive an expression for θ_r in terms of θ_i and e. [Ans: $\tan^{-1}(e \tan \theta_i)$.]

(b) Derive an expression for v' in terms of v, θ_i, and e. [Ans: $\sqrt{e^2 + (1 - e^2) \cos^2 \theta_i}\, v$.]

6.10 Consider an elastic collision between two equal-mass objects. After the collision, the initially moving object moves off at an angle θ_1 to its initial direction of motion, whereas the initially stationary object recoils at an angle θ_2 to this direction. See Figure 6.12.

(a) Demonstrate that, after the collision, the two objects move off at right-angles to one another; that is, $\theta_1 + \theta_2 = \pi/2$.

(b) Suppose that the two objects are smooth spheres, and let μ be their impact parameter, as defined in Equation (6.114). Show that

$$\theta_1 = \arccos \mu,$$

$$\theta_2 = \frac{\pi}{2} - \arccos \mu.$$

6.11 Two particles of masses m_1 and m_2, moving with speeds v_1 and v_2, respectively, in directions inclined to one another at an angle θ, collide and coalesce. Demonstrate that the particles' subsequent speed is

$$\frac{\sqrt{(m_1^2\, v_1^2 + 2\, m_1\, m_2\, v_1\, v_2\, \cos\theta + m_2^2\, v_2^2}}{m_1 + m_2},$$

and that the loss of energy is

$$\frac{m_1\, m_2}{2\,(m_1 + m_2)}\,(v_1^2 - 2\, v_1\, v_2\, \cos\theta + v_2^2).$$

CHAPTER 7

Circular Motion

7.1 INTRODUCTION

Up to now, we have essentially only considered rectilinear motion; that is, motion in a straight-line. Let us now broaden our approach so as to take into account the most important type of non-rectilinear motion; namely, **circular motion**.

7.2 UNIFORM CIRCULAR MOTION

Suppose that an object executes a circular orbit of radius r with the uniform tangential speed v. The instantaneous position of the object is most conveniently specified in terms of an angle θ. See Figure 7.1. For instance, we could decide that $\theta = 0°$ corresponds to the object's location at $t = 0$, in which case we would write

$$\theta(t) = \omega\,t, \tag{7.1}$$

where ω is termed the *angular velocity* of the object. For a uniformly rotating object, the angular velocity is simply the angle through which the object (or to be more exact, the radius vector connecting the object to the center of the circle) turns in one second.

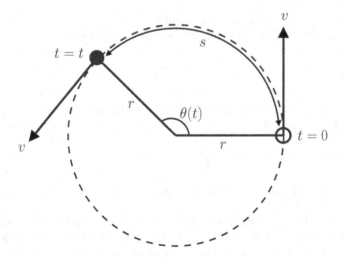

Figure 7.1 Circular motion

Consider the motion of the object in the time interval between $t = 0$ and $t = t$. In this interval, the object rotates through an angle θ, and traces out a circular arc of length s. See

DOI: 10.1201/9781003198642-7

Figure 7.1. It is fairly obvious that the arc-length, s, is directly proportional to the angle, θ. But, what is the constant of proportionality? Clearly, an angle of $360°$ corresponds to an arc-length of $2\pi r$ (i.e., the circumference of the circle). Hence, an angle θ must correspond to an arc-length of

$$s = 2\pi r \frac{\theta(°)}{360°}. \tag{7.2}$$

At this stage, it is convenient to define a new angular unit known as a *radian* (symbol rad). An angle measured in radians is related to an angle measured in degrees via the following simple formula:

$$\theta(\text{rad}) = \frac{2\pi}{360°} \theta(°). \tag{7.3}$$

Thus, $360°$ corresponds to 2π radians, $180°$ corresponds to π radians, $90°$ corresponds to $\pi/2$ radians, and $57.296°$ corresponds to 1 radian. When θ is measured in radians, Equation (7.2) simplifies greatly to give

$$s = r\,\theta. \tag{7.4}$$

Henceforth, in this book, all angles are measured in radians by default.

Consider the motion of the object in the short interval between times t and $t + \delta t$. In this interval, the object turns through a small angle $\delta\theta$, and traces out a short arc of length δs, where

$$\delta s = r\,\delta\theta. \tag{7.5}$$

Now, $\delta s/\delta t$ (i.e., distance moved per unit time) is the object's tangential velocity, v, whereas $\delta\theta/\delta t$ (i.e., angle turned through per unit time) is the object's angular velocity, ω. Thus, dividing Equation (7.5) by δt, we obtain

$$v = r\,\omega. \tag{7.6}$$

Note, however, that this relation is only valid if the angular velocity, ω, is measured in radians per second. From now on, in this book, all angular velocities are measured in radians per second by default.

An object that rotates with uniform angular velocity ω turns through ω radians in 1 second. Hence, the object turns through 2π radians (i.e., it executes a complete circle) in

$$T = \frac{2\pi}{\omega} \tag{7.7}$$

seconds. Here, T is the repetition period of the circular motion. If the object executes a complete cycle (i.e., turns through $360°$) in T seconds then the number of cycles executed per second is

$$f = \frac{1}{T} = \frac{\omega}{2\pi}. \tag{7.8}$$

Here, the repetition frequency, f, of the motion is measured in cycles per second; otherwise known as *hertz* (symbol Hz).

As an example, suppose that an object executes uniform circular motion, of radius $r = 1.2\,\text{m}$, at a frequency of $f = 50\,\text{Hz}$ (i.e., the object executes a complete rotation 50 times a second). The repetition period of this motion is simply

$$T = \frac{1}{f} = 0.02\,\text{s}. \tag{7.9}$$

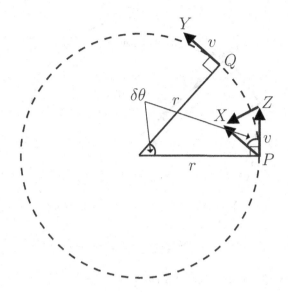

Figure 7.2 Centripetal acceleration

Furthermore, the angular frequency of the motion is given by

$$\omega = 2\pi\, f = 314.16\,\text{rad/s}. \tag{7.10}$$

Finally, the tangential velocity of the object is

$$v = r\,\omega = 1.2 \times 314.16 = 376.99\,\text{m/s}. \tag{7.11}$$

7.3 CENTRIPETAL ACCELERATION

An object executing a circular orbit of radius r with uniform tangential speed v possesses a velocity vector, $\mathbf{v}$, whose magnitude is constant, but whose direction is continually changing. It follows that the object must be accelerating, because (vector) acceleration is the rate of change of (vector) velocity, and the (vector) velocity is varying in time.

Suppose that the object moves from point P to point Q between times t and $t + \delta t$, as shown in Figure 7.2. Suppose, further, that the object rotates through $\delta\theta$ radians in this time interval. The vector $\overrightarrow{PX}$, shown in the diagram, is identical to the vector $\overrightarrow{QY}$ (which is the object's velocity vector at time $t + \delta t$). Moreover, the angle subtended between vectors $\overrightarrow{PZ}$ (which is the object's velocity vector at time t) and $\overrightarrow{PX}$ is $\delta\theta$. The vector $\overrightarrow{ZX}$ represents the change in vector velocity, $\delta\mathbf{v}$, between times t and $t + \delta t$. It can be seen that this vector is directed toward the center of the circle. From standard trigonometry, the length of vector $\overrightarrow{ZX}$ is

$$\delta v = 2\,v\,\sin(\delta\theta/2), \tag{7.12}$$

because vectors $\overrightarrow{PZ}$ and $\overrightarrow{PX}$ are both of length v, and the angle subtended between them is $\delta\theta$. However, for small angles, $\sin\theta \simeq \theta$, provided that θ is measured in radians. Hence,

$$\delta v \simeq v\,\delta\theta. \tag{7.13}$$

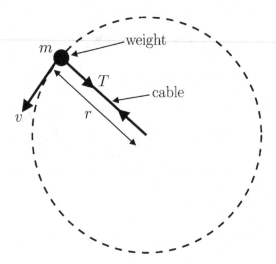

Figure 7.3 Rotating weight on the end of a cable

It follows that the object's acceleration is

$$a = \frac{\delta v}{\delta t} = v\,\frac{\delta \theta}{\delta t} = v\,\omega, \tag{7.14}$$

where $\omega = \delta\theta/\delta t$ is the angular velocity of the object, measured in radians per second. In summary, an object executing a circular orbit, of radius r, with uniform tangential velocity v, and uniform angular velocity $\omega = v/r$, possesses an acceleration directed toward the center of the circle—that is, a *centripetal* acceleration—of magnitude

$$a = v\,\omega = \frac{v^2}{r} = r\,\omega^2. \tag{7.15}$$

7.4   ROTATING WEIGHT ON THE END OF A CABLE

Suppose that a weight, of mass m, is attached to the end of a cable, of length r, and whirled around such that the weight executes a horizontal circle, of radius r, with uniform tangential velocity, v. (See Figure 7.3.) As we have just seen, the weight is subject to a centripetal acceleration of magnitude v^2/r. Hence, the weight experiences a centripetal force

$$f = \frac{m\,v^2}{r}. \tag{7.16}$$

What provides this force? In the present example, the force is provided by the tension, T, in the cable. Hence,

$$T = \frac{m\,v^2}{r}. \tag{7.17}$$

Suppose that the cable is such that it will snap if its tension exceeds a certain critical value, $T_{\max}$. It follows that there is a maximum velocity with which the weight can be whirled around; namely,

$$v_{\max} = \sqrt{\frac{r\,T_{\max}}{m}}. \tag{7.18}$$

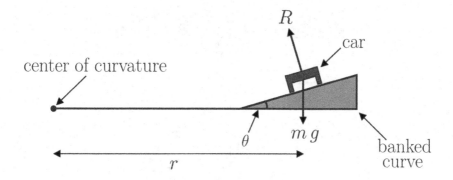

Figure 7.4 A banked curve

If v exceeds v_{max} then the cable will snap. As soon as the cable snaps, the weight will cease to be subject to a centripetal force, so it will fly off—with velocity v_{max}—along the straight-line that is tangential to the circular orbit that it was previously executing.

7.5 BANKED CURVE

Civil engineers generally bank curves on roads in such a manner that a car attempting the curve at the recommended speed does not have to rely on friction between its tires and the road surface in order to get around round the curve. Let us examine the physics behind banked curves.

Let r be the radius of curvature of the curve in question. Let θ be the banking angle. Let v be the recommended speed for going around the curve. See Figure 7.4. Consider a car of mass m going around the curve at the recommended speed. The car's weight, $m\,g$, acts vertically downward. The road surface exerts an upward normal reaction, R, on the car. The vertical component of the reaction must balance the downward weight of the car, so

$$R\cos\theta = m\,g. \tag{7.19}$$

The horizontal component of the reaction, $R\sin\theta$, acts toward the center of curvature of the road; this component provides the force $m\,v^2/r$ acting toward the center of the curvature that the car experiences as it rounds the curve. In other words,

$$R\sin\theta = \frac{m\,v^2}{r}. \tag{7.20}$$

The previous two equations can be combined to give

$$\tan\theta = \frac{v^2}{r\,g}. \tag{7.21}$$

Hence, the optimum banking angle is

$$\theta = \tan^{-1}\left(\frac{v^2}{r\,g}\right). \tag{7.22}$$

Note that if the car attempts to go around the curve at the wrong speed then $R\sin\theta \neq m\,v^2/r$, and the difference has to be made up by a sideways friction force exerted between the car's tires and the road surface. Unfortunately, this does not always work; especially if the road surface is wet.

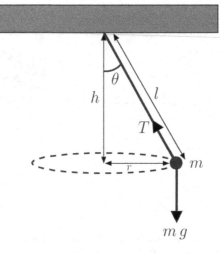

Figure 7.5 A conical pendulum

7.6 CONICAL PENDULUM

Suppose that an object, of mass m, is attached to the end of a light, inextensible cable whose other end is attached to a rigid beam. Suppose, further, that the object is given an initial horizontal velocity such that it executes a horizontal circular orbit of radius r with angular velocity ω. See Figure 7.5. Let h be the vertical distance between the beam and the plane of the circular orbit, and let θ be the angle subtended by the cable with the downward vertical. This dynamic system is known as a *conical pendulum* because the cable traces out a cone as the object rotates.

The object is subject to two forces; a gravitational force, $m\,g$, that acts vertically downward, and a tension force, T, that acts upward along the cable. The tension force can be resolved into a component $T\cos\theta$ that acts vertically upward, and a component $T\sin\theta$ that acts toward the center of the circle. Force balance in the vertical direction yields

$$T\cos\theta = m\,g. \tag{7.23}$$

In other words, the vertical component of the tension force balances the weight of the object.

Because the object is executing a circular orbit, of radius r, with angular velocity ω, it experiences a centripetal acceleration $\omega^2 r$. Hence, it is subject to a centripetal force $m\,\omega^2 r$. This force is provided by the component of the cable tension that acts toward the center of the circle. In other words,

$$T\sin\theta = m\,\omega^2 r. \tag{7.24}$$

Taking the ratio of Equations (7.23) and (7.24), we obtain

$$\tan\theta = \frac{\omega^2 r}{g}. \tag{7.25}$$

However, by simple trigonometry,

$$\tan\theta = \frac{r}{h}. \tag{7.26}$$

Hence, we find that

$$\omega = \sqrt{\frac{g}{h}}. \tag{7.27}$$

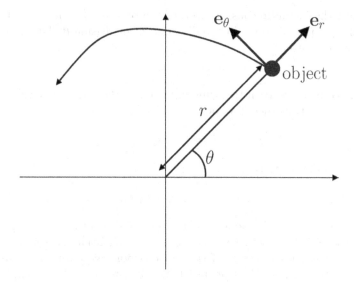

Figure 7.6 Polar coordinates

Note that if l is the length of the cable then $h = l \cos\theta$. It follows that

$$\omega = \sqrt{\frac{g}{l \cos\theta}}. \tag{7.28}$$

For instance, if the length of the cable is $l = 0.2\,\mathrm{m}$ and the conical angle is $\theta = 30°$, then the angular velocity of rotation is given by

$$\omega = \sqrt{\frac{9.81}{0.2 \times \cos 30°}} = 7.526\,\mathrm{rad/s}, \tag{7.29}$$

which translates to a rotation frequency in cycles per second of

$$f = \frac{\omega}{2\pi} = 1.20\,\mathrm{Hz}. \tag{7.30}$$

7.7 NON-UNIFORM CIRCULAR MOTION

Consider an object that executes non-uniform circular motion, as illustrated in Figure 7.6. Suppose that the motion is confined to a two-dimensional plane. We can specify the instantaneous position of the object in terms of its polar coordinates, r and θ. Here, r is the radial distance of the object from the origin of our coordinate system, whereas θ is the angular bearing of the object from the origin, measured with respect to some arbitrarily chosen direction. In general, both r and θ are changing in time. As an example of non-uniform circular motion, consider the motion of the Earth around the Sun. Suppose that the origin of our coordinate system corresponds to the position of the Sun. As the Earth rotates, its angular bearing, θ, relative to the Sun, obviously changes in time. However, because the Earth's orbit is slightly elliptical, its radial distance, r, from the Sun also varies in time. Moreover, as the Earth moves closer to the Sun, its rate of rotation speeds up, and vice versa. Hence, the rate of change of θ with time is non-uniform.

Let us define two unit vectors, $\mathbf{e}_r$ and $\mathbf{e}_\theta$. Incidentally, a unit vector simply a vector whose length is unity. As shown in Figure 7.6, the radial unit vector, $\mathbf{e}_r$, always points from the origin toward the instantaneous position of the object. Moreover, the tangential unit vector, $\mathbf{e}_\theta$, is always normal to $\mathbf{e}_r$, in the direction of increasing θ. The position vector, $\mathbf{r}$, of the object can be written

$$\mathbf{r} = r\,\mathbf{e}_r. \tag{7.31}$$

In other words, vector $\mathbf{r}$ points in the same direction as the radial unit vector $\mathbf{e}_r$, and is of length r. We can write the object's velocity in the form

$$\mathbf{v} \equiv \dot{\mathbf{r}} = v_r\,\mathbf{e}_r + v_\theta\,\mathbf{e}_\theta, \tag{7.32}$$

whereas the acceleration is written

$$\mathbf{a} \equiv \dot{\mathbf{v}} = a_r\,\mathbf{e}_r + a_\theta\,\mathbf{e}_\theta. \tag{7.33}$$

Here, v_r is termed the object's radial velocity, while v_θ is termed its tangential velocity. Likewise, a_r is the radial acceleration, and a_θ is the tangential acceleration. But, how do we express these quantities in terms of the object's polar coordinates, r and θ? It turns out that this is a far from straightforward task. For instance, if we simply differentiate Equation (7.31) with respect to time, we obtain

$$\mathbf{v} = \dot{r}\,\mathbf{e}_r + r\,\dot{\mathbf{e}}_r, \tag{7.34}$$

where $\dot{\mathbf{e}}_r$ is the time derivative of the radial unit vector; this quantity is non-zero because $\mathbf{e}_r$ changes direction as the object moves. Unfortunately, it is not entirely clear how to evaluate $\dot{\mathbf{e}}_r$. In the following, we outline a famous trick for calculating v_r, v_θ, etcetera, without ever having to evaluate the time derivatives of the unit vectors $\mathbf{e}_r$ and $\mathbf{e}_\theta$.

Consider a general complex number,

$$z = x + \mathrm{i}\,y, \tag{7.35}$$

where x and y are real numbers, and i is the square root of -1 (i.e., $\mathrm{i}^2 = -1$). Here, x is the real part of z, whereas y is the imaginary part. We can visualize z as a point in the so-called *complex plane*; that is, a two-dimensional plane in which the real parts of complex numbers are plotted along one Cartesian axis, whereas the corresponding imaginary parts are plotted along the other axis. Thus, the coordinates of z in the complex plane are simply $(x,\ y)$. See Figure 7.7. In other words, we can use a complex number to represent a position vector in a two-dimensional plane. Note that the length of the vector is equal to the modulus of the corresponding complex number. Incidentally, the modulus of $z = x + \mathrm{i}\,y$ is defined

$$|z| = \sqrt{x^2 + y^2}. \tag{7.36}$$

Consider the complex number $\mathrm{e}^{\mathrm{i}\theta}$, where θ is real. A famous result in complex analysis—known as *Euler's theorem*—allows us to split this number into its real and imaginary components:

$$\mathrm{e}^{\mathrm{i}\theta} = \cos\theta + \mathrm{i}\,\sin\theta. \tag{7.37}$$

Now, as we have just discussed, we can think of $\mathrm{e}^{\mathrm{i}\theta}$ as representing a vector in the complex plane; the real and imaginary parts of $\mathrm{e}^{\mathrm{i}\theta}$ form the coordinates of the head of the vector, whereas the tail of the vector corresponds to the origin. What are the properties of this vector? The length of the vector is given by

$$\left|\mathrm{e}^{\mathrm{i}\theta}\right| = \sqrt{\cos^2\theta + \sin^2\theta} = 1. \tag{7.38}$$

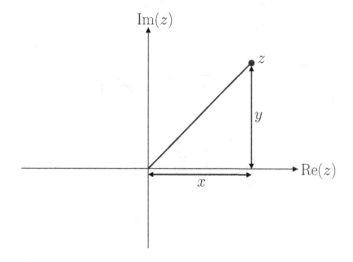

Figure 7.7   Representation of a complex number in the complex plane

In other words, $e^{i\theta}$ represents a unit vector. In fact, it is clear from Figure 7.8 that $e^{i\theta}$ represents the radial unit vector, $\mathbf{e}_r$, for an object whose angular polar coordinate (measured anti-clockwise from the real axis) is θ. Can we also find a complex representation of the corresponding tangential unit vector, $\mathbf{e}_\theta$? Actually, we can. The complex number $i\,e^{i\theta}$ can be written

$$i\,e^{i\theta} = -\sin\theta + i\,\cos\theta. \tag{7.39}$$

Here, we have just multiplied Equation (7.37) by i, making use of the fact that $i^2 = -1$. This complex number again represents a unit vector, because

$$\left| i\,e^{i\theta} \right| = \sqrt{\sin^2\theta + \cos^2\theta} = 1. \tag{7.40}$$

Moreover, as is clear from Figure 7.8, this vector is normal to $\mathbf{e}_r$, in the direction of increasing θ. In other words, $i\,e^{i\theta}$ represents the tangential unit vector, $\mathbf{e}_\theta$.

Consider an object executing non-uniform circular motion in the complex plane. By analogy with Equation (7.31), we can represent the instantaneous position vector of this object via the complex number

$$z = r\,e^{i\theta}. \tag{7.41}$$

Here, $r(t)$ is the object's radial distance from the origin, whereas $\theta(t)$ is its angular bearing relative to the real axis. Note that, in the previous formula, we are using $e^{i\theta}$ to represent the radial unit vector, $\mathbf{e}_r$. Now, if z represents the position vector of the object, then $\dot{z} = dz/dt$ must represent the object's velocity vector. Differentiating Equation (7.41) with respect to time, using the standard rules of calculus, we obtain

$$\dot{z} = \dot{r}\,e^{i\theta} + r\,\dot{\theta}\,i\,e^{i\theta}. \tag{7.42}$$

Comparing with Equation (7.32), recalling that $e^{i\theta}$ represents $\mathbf{e}_r$, and $i\,e^{i\theta}$ represents $\mathbf{e}_\theta$, we obtain

$$v_r = \dot{r}, \tag{7.43}$$

$$v_\theta = r\,\dot{\theta} = r\,\omega, \tag{7.44}$$

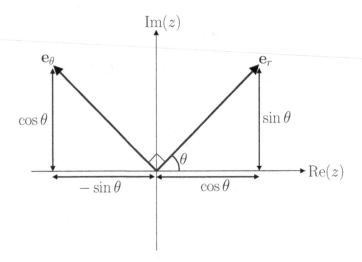

Figure 7.8   Representation of the unit vectors e_r and e_θ in the complex plane

where $\omega = d\theta/dt$ is the object's instantaneous angular velocity. Thus, as desired, we have obtained expressions for the radial and tangential velocities of the object in terms of its polar coordinates, r and θ. We can go further. Let us differentiate $\dot{z}$ with respect to time, in order to obtain a complex number representing the object's vector acceleration. Again, using the standard rules of calculus, we obtain

$$\ddot{z} = (\ddot{r} - r\,\dot{\theta}^2)\,e^{i\theta} + (r\,\ddot{\theta} + 2\,\dot{r}\,\dot{\theta})\,i\,e^{i\theta}. \tag{7.45}$$

Comparing with Equation (7.33), recalling that $e^{i\theta}$ represents e_r and $i\,e^{i\theta}$ represents e_θ, we obtain

$$a_r = \ddot{r} - r\,\dot{\theta}^2 = \ddot{r} - r\,\omega^2, \tag{7.46}$$

$$a_\theta = r\,\ddot{\theta} + 2\,\dot{r}\,\dot{\theta} = r\,\dot{\omega} + 2\,\dot{r}\,\omega. \tag{7.47}$$

Thus, we now have expressions for the object's radial and tangential accelerations in terms of r and θ. The beauty of this derivation is that the complex analysis has automatically taken care of the fact that the unit vectors e_r and e_θ change direction as the object moves.

Let us now consider the commonly occurring special case in which an object executes a circular orbit at fixed radius, but varying angular velocity. Because the radius is fixed, it follows that $\dot{r} = \ddot{r} = 0$. According to Equations (7.43) and (7.44), the radial velocity of the object is zero, and the tangential velocity takes the form

$$v_\theta = r\,\omega. \tag{7.48}$$

Note that the previous equation is exactly the same as Equation (7.6); the only difference is that we have now proved that this relation holds for non-uniform, as well as uniform, circular motion. According to Equation (7.46), the radial acceleration is given by

$$a_r = -r\,\omega^2. \tag{7.49}$$

The minus sign indicates that this acceleration is directed toward the center of the circle. Of course, the previous equation is equivalent to Equation (7.15); the only difference is that

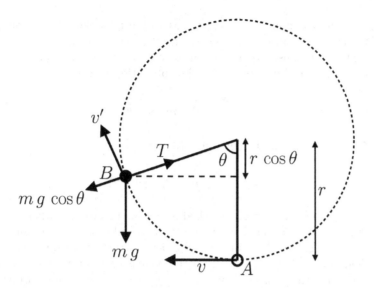

Figure 7.9 Motion in a vertical circle

we have now proved that this relation holds for non-uniform, as well as uniform, circular motion. Finally, according to Equation (7.47), the tangential acceleration takes the form

$$a_\theta = r\,\dot{\omega}.\qquad(7.50)$$

The existence of a non-zero tangential acceleration (in the former case) is the one difference between non-uniform and uniform circular motion (at constant radius).

7.8 VERTICAL PENDULUM

Let us now examine an example of non-uniform circular motion. Suppose that an object of mass m is attached to the end of a light rigid rod, or light cable, of length r. The other end of the rod, or cable, is attached to a stationary pivot in such a manner that the object is free to execute a vertical circle about this pivot. Let θ measure the angular position of the object, measured with respect to the downward vertical. Let v be the velocity of the object at $\theta = 0$. How large do we have to make v in order for the object to execute a complete vertical circle?

Consider Figure 7.9. Suppose that the object moves from point A, where its tangential velocity is v, to point B, where its tangential velocity is v'. Let us, first of all, obtain the relationship between v and v'. This is most easily achieved by considering energy conservation. At point A, the object is situated a vertical distance r below the pivot, whereas at point B the vertical distance below the pivot has been reduced to $r\cos\theta$. Hence, in moving from A to B, the object gains potential energy $m\,g\,r\,(1-\cos\theta)$. This gain in potential energy must be offset by a corresponding loss in kinetic energy. Thus,

$$\frac{1}{2}\,m\,v^2 - \frac{1}{2}\,m\,v'^2 = m\,g\,r\,(1-\cos\theta),\qquad(7.51)$$

which reduces to

$$v'^2 = v^2 - 2\,r\,g\,(1-\cos\theta).\qquad(7.52)$$

Let us now examine the radial acceleration of the object at point B. The radial forces acting on the object are the tension, T, in the rod, or cable, which acts toward the center of the circle, and the component $m g \cos \theta$ of the object's weight, which acts away from the center of the circle. Because the object is executing circular motion with instantaneous tangential velocity v', it must experience an instantaneous acceleration v'^2/r toward the center of the circle. Hence, Newton's second law of motion yields

$$\frac{m v'^2}{r} = T - m g \cos \theta. \tag{7.53}$$

Equations (7.52) and (7.53) can be combined to give

$$T = \frac{m v^2}{r} + m g \, (3 \, \cos \theta - 2). \tag{7.54}$$

Suppose that the object is, in fact, attached to the end of a cable, rather than a rigid rod. One important property of cables is that, unlike rigid rods, they cannot support negative tensions. In other words, a cable can only pull objects attached to its two ends together; it cannot push them apart. Another way of putting this is that if the tension in a cable ever becomes negative then the cable will become slack and collapse. Thus, if our object is to execute a full vertical circle then the tension, T, in the cable must remain positive for all values of θ. It is clear from Equation (7.54) that the tension attains its minimum value when $\theta = \pi$ (at which point $\cos \theta = -1$). This is hardly surprising, because $\theta = \pi$ corresponds to the point at which the object attains its maximum height, and, therefore, its minimum tangential velocity. It is certainly the case that if the cable tension is positive at this point then it must be positive at all other points. Now, the tension at $\theta = \pi$ is given by

$$T_0 = \frac{m v^2}{r} - 5 \, m \, g. \tag{7.55}$$

Hence, the condition for the object to execute a complete vertical circle without the cable becoming slack is $T_0 > 0$, or

$$v^2 > 5 \, r \, g. \tag{7.56}$$

Note that this condition is independent of the mass of the object.

Suppose that the object is attached to the end of a rigid rod, instead of a cable. There is now no constraint on the tension, because a rigid rod can quite easily support a negative tension (i.e., it can push, as well as pull, on objects attached to its two ends). However, in order for the object to execute a complete vertical circle, the square of its tangential velocity v'^2 must remain positive at all values of θ. It is clear from Equation (7.52) that v'^2 attains its minimum value when $\theta = \pi$. This is, again, hardly surprising. Thus, if v'^2 is positive at this point then it must be positive at all other points. Now, the expression for v'^2 at $\theta = \pi$ is

$$v_0'^2 = v^2 - 4 \, r \, g. \tag{7.57}$$

Hence, the condition for the object to execute a complete vertical circle is $v_0'^2 > 0$, or

$$v^2 > 4 \, r \, g. \tag{7.58}$$

Note that this condition is slightly easier to satisfy than the condition (7.56). In other words, it is slightly easier to cause an object attached to the end of a rigid rod to execute a vertical circle than it is to cause an object attached to the end of a cable to execute the same circle. The reason for this is that the rigidity of the rod helps support the object when it is situated above the pivot point.

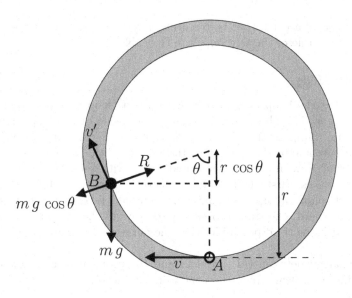

Figure 7.10 Motion on the inside of a vertical hoop

7.9 MOTION ON CURVED SURFACES

Consider a smooth, rigid, vertical hoop of internal radius r, as shown in Figure 7.10. Suppose that an object of mass m slides without friction around the inside of this hoop. What is the motion of this object? Is it possible for the object to execute a complete vertical circle?

Suppose that the object moves from point A to point B in Figure 7.10. In doing so, it gains potential energy $mgr(1 - \cos\theta)$, where θ is the angular coordinate of the object measured with respect to the downward vertical. This gain in potential energy must be offset by a corresponding loss in kinetic energy. Thus,

$$\frac{1}{2}mv^2 - \frac{1}{2}mv'^2 = mgr(1 - \cos\theta), \qquad (7.59)$$

which reduces to

$$v'^2 = v^2 - 2rg(1 - \cos\theta). \qquad (7.60)$$

Here, v is the velocity at point A ($\theta = 0$) and v' is the velocity at point B ($\theta = \theta$).

Let us now examine the radial acceleration of the object at point B. The radial forces acting on the object are the reaction, R, of the vertical hoop, which act toward the center of the hoop, and the component $mg\cos\theta$ of the object's weight, which acts away from the center of the hoop. Because the object is executing circular motion with instantaneous tangential velocity v', it must experience an instantaneous acceleration v'^2/r toward the center of the hoop. Hence, Newton's second law of motion yields

$$\frac{mv'^2}{r} = R - mg\cos\theta. \qquad (7.61)$$

Note, however, that there is a constraint on the reaction, R, that the hoop can exert on the object; this reaction must always be positive. In other words, the hoop can push the object away from itself, but it can never pull it toward itself. Another way of putting this is that if the reaction ever becomes negative then the object will fly off the surface of the hoop, because it is no longer being pressed into this surface. It should be clear, by now, that the problem we are considering is exactly analogous to the earlier problem of an object attached to the end of a cable that is executing a vertical circle, with the reaction, R, of the hoop playing the role of the tension, T, in the cable.

7.9.1 Fairground Ride

Let us imagine that the hoop under consideration is a "loop the loop" segment in a fairground roller-coaster. The object sliding around the inside of the loop then becomes the roller-coaster train. Suppose that the fairground operator can vary the velocity, v, with which the train is sent into the bottom of the loop (i.e., the velocity at $\theta = 0$). What is the safe range of v? Now, if the train starts at $\theta = 0$ with velocity v, then there are only three possible outcomes. First, the train can execute a complete circuit of the loop. Second, the train can slide part way up the loop, come to a halt, reverse direction, and then slide back down again. Third, the train can slide part way up the loop, but then falls off the loop. Obviously, it is the third possibility that the fairground operator would wish to guard against.

Using the analogy between this problem and the problem of a weight on the end of a cable executing a vertical circle, the condition for the roller-coaster train to execute a complete circuit is

$$v^2 > 5\,r\,g. \tag{7.62}$$

Note, interestingly enough, that this condition is independent of the mass of the train.

Equation (7.61) yields

$$v'^2 = \frac{r\,R}{m} - r\,g\,\cos\theta. \tag{7.63}$$

Now, the condition for the train to reverse direction without falling off the loop is $v'^2 = 0$ with $R > 0$. Thus, the train reverses direction when

$$R = m\,g\,\cos\theta. \tag{7.64}$$

Note that this equation can only be satisfied for positive R when $\cos\theta > 0$. In other words, the train can only turn around without falling off the loop if the turning point lies in the lower half of the loop (i.e., $-\pi/2 < \theta < \pi/2$). The condition for the train to fall off the loop is $R = 0$, or

$$v'^2 = -r\,g\,\cos\theta. \tag{7.65}$$

Note that this equation can only be satisfied for positive v'^2 when $\cos\theta < 0$. In other words, the train can only fall off the loop when it is situated in the upper half of the loop. It is fairly clear that if the train's initial velocity is not sufficiently large for it to execute a complete circuit of the loop, and not sufficiently small for it to turn around before entering the upper half of the loop, then it must inevitably fall off the loop somewhere in the loop's upper half. The critical value of v^2 above which the train executes a complete circuit is $5\,r\,g$. [See Equation (7.62).] The critical value of v^2 at which the train just turns around before entering the upper half of the loop is $2\,r\,g$. [This is obtained from Equation (7.60) by setting $v' = 0$ and $\theta = \pi/2$.] Hence, the dangerous range of v^2 is

$$2\,r\,g < v^2 < 5\,r\,g. \tag{7.66}$$

For $v^2 < 2\,r\,g$, the train turns around in the lower half of the loop. For $v^2 > 5\,r\,g$, the train executes a complete circuit around the loop. However, for $2\,r\,g < v^2 < 5\,r\,g$, the train falls off the loop somewhere in its upper half.

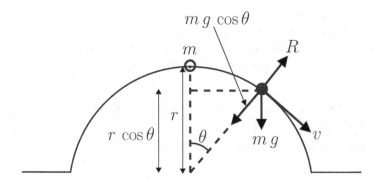

7.11 A skier on a hemispherical mountain.

Skier on a Hemispherical Mountain

ider a skier of mass m skiing down a hemispherical mountain of radius r, as shown gure 7.11. Let θ be the angular coordinate of the skier, measured with respect to the .rd vertical. Suppose that the skier starts at rest ($v = 0$) on top of the mountain 0), and slides down the mountain without friction. At what point does the skier fly off urface of the mountain?

uppose that the skier has reached angular coordinate θ. At this stage, the skier has a though a height $r(1 - \cos\theta)$. Thus, the tangential velocity, v, of the skier is given by y conservation:

$$\frac{1}{2}mv^2 = mgr(1 - \cos\theta). \tag{7.67}$$

is now consider the skier's radial acceleration. The radial forces acting on the skier are eaction, R, exerted by the mountain, which acts radially outward, and the component e skier's weight $mg\cos\theta$, which acts radially inward. Because the skier is executing lar motion of radius r, with instantaneous tangential velocity v, he/she experiences an ntaneous inward radial acceleration v^2/r. Hence, Newton's second law of motion yields

$$\frac{mv^2}{r} = mg\cos\theta - R. \tag{7.68}$$

quations (7.67) and (7.68) can be combined to give

$$R = mg(3\cos\theta - 2). \tag{7.69}$$

efore, the reaction R is constrained to be positive; the mountain can push outward e skier, but it cannot pull the skier inward. In fact, as soon as the reaction becomes tive, the skier flies of the surface of the mountain. This occurs when $\cos\theta_0 = 2/3$, $= 48.19°$. The height through which the skier falls before becoming a ski-jumper is $r(1 - \cos\theta_0) = a/3$.

EXERCISES

1 A car of mass $m = 2000\,\text{kg}$ travels around a flat circular race track of radius $r = 85\,\text{m}$. The car starts at rest, and its speed increases at the constant rate $a_\theta = 0.6\,\text{m/s}^2$.

7.2 An amusement park ride consists of a vertical cylinder that spins about a vertical axis. When the cylinder spins sufficiently fast, any person inside it is held up against the wall. Suppose that the coefficient of static friction between a typical person and the wall is $\mu = 0.25$. Let the mass of an typical person be $m = 60\,\text{kg}$, and let $r = 7\,\text{m}$ be the radius of the cylinder.

   (a) What is the critical angular velocity of the cylinder above which a typical person will not slide down the wall? [Ans: $2.37\,\text{rad/s}$.]

   (b) How many revolutions per second is the cylinder executing at this critical velocity? [Ans: $0.38\,\text{Hz}$.]

7.3 A small weight whose mass is $25\,\text{g}$ is whirled in a horizontal circular path on the end of a cable of length $400\,\text{mm}$. What is the tension in the cable when the weight's angular speed of rotation is $30\,\text{rad/s}$? [Ans: $9.0\,\text{N}$.]

7.4 A stunt pilot experiences weightlessness momentarily at the top of a "loop the loop" maneuver (i.e., flying in a vertical circle). Given that the speed of the stunt plane is $v = 500\,\text{km/h}$, what is the radius r of the loop? [Ans: $1.97\,\text{km}$.]

7.5 A bullet of mass $m = 10\,\text{g}$ strikes a pendulum bob of mass $M = 1.3\,\text{kg}$ horizontally with speed v, and then becomes embedded in the bob. The bob is initially at rest, and is suspended by a stiff rod of length $l = 0.6\,\text{m}$ and negligible mass. The bob is free to rotate in the vertical direction.

   (a) What is the minimum value of v that causes the bob to execute a complete vertical circle? [Ans: $635.6\,\text{m/s}$.]

   (b) How does the answer change if the bob is suspended from a light flexible cable (of the same length), instead of a stiff rod? [Ans: $710.7\,\text{m/s}$.]

7.6 Consider a banked curve of banking angle θ and radius of curvature r. Suppose that a vehicle goes around the curve at speed v. Let μ be the coefficient of friction between the vehicle and the road surface.

   (a) What is the maximum speed that the vehicle can go around the curve without skidding? [Ans: $\sqrt{rg}\,(\sin\theta + \mu\,\cos\theta)^{1/2}/(\cos\theta - \mu\sin\theta)^{1/2}$ if $\tan\theta < 1/\mu$, and ∞ if $\tan\theta > 1/\mu$.]

   (b) What is the minimum speed that the vehicle can go around the curve without skidding? [Ans: $\sqrt{rg}\,(\sin\theta - \mu\,\cos\theta)^{1/2}/(\cos\theta + \mu\sin\theta)^{1/2}$ if $\tan\theta > \mu$, and 0 if $\tan\theta < \mu$.]

7.7 A weight is suspended by two equal cables from two points at the same level. The inclination of each cable to the vertical is α. If one cable is severed, show that the tension in the other is instantaneously modified by a factor $2\cos^2\alpha$. [From Lamb 1942.]

7.8 A weight is suspended by two cables whose inclinations to the vertical are α and β. Show that if the second cable is severed then the tension in the first is instantaneously modified by a factor

$$\frac{\sin\beta}{\sin(\alpha+\beta)\,\cos\alpha}.$$

A simple pendulum is started so as to make complete revolutions in a vertical plane.

(a) If ω_1 and ω_2 are the greatest and least angular velocities, show that the angular velocity when the pendulum subtends an angle θ with the downward vertical is

$$\sqrt{\omega_1^2 \, \cos^2(\theta/2) + \omega_2^2 \, \sin^2(\theta/2)}.$$

(b) If T_1 and T_2 are the greatest and least tensions in the pendulum cable, show that the tension when the pendulum subtends an angle θ with the downward vertical is

$$T_1 \, \cos^2(\theta/2) + T_2 \, \sin^2(\theta/2).$$

[From Lamb 1942.]

tational Motion

INTRODUCTION

now, we have only analyzed the dynamics of point masses (i.e., objects whose spatial
is either negligible, or plays no role in their motion). Let us now broaden our approach
er to take extended objects into account. The only type of motion that a point mass
can exhibit is translational motion; that is, motion by which the object moves from
oint in space to another. However, an extended object can exhibit another, quite
ct, type of motion by which it remains located (more or less) at the same spatial
on, but constantly changes its orientation with respect to other fixed points in space.
new type of motion is called **rotation**. Let us investigate rotational motion.

RIGID BODY ROTATION

der a rigid body executing pure rotational motion (i.e., rotational motion that has no
ational component). It is possible to define an *axis of rotation* (which, for the sake of
city, is assumed to pass through the body); this axis corresponds to the straight-line
s the locus of all points inside the body that remain stationary as the body rotates.
eral point located inside the body executes circular motion that is centered on the
on axis, and orientated in the plane perpendicular to this axis. In the following, we
acitly assume that the axis of rotation remains fixed.
gure 8.1 shows a typical rigidly rotating body. The axis of rotation is the line AB. A
al point, P, lying within the body executes a circular orbit, centered on AB, in the
perpendicular to AB. Let the line QP be a radius of this orbit that links the axis of
on to the instantaneous position of P at time t. Obviously, this implies that QP is
l to AB. Suppose that at time $t + \delta t$ point P has moved to P', and the radius QP
tated through an angle $\delta\phi$. The instantaneous *angular velocity* of the body, $\omega(t)$, is
d

$$\omega \equiv \lim_{\delta t \to 0} \frac{\delta\phi}{\delta t} = \frac{d\phi}{dt}. \tag{8.1}$$

that if the body is indeed rotating rigidly then the calculated value of ω should be the
for all possible points P lying within the body (except for those points lying exactly on
xis of rotation, for which ω is ill-defined). The rotation speed, v, of point P is related
angular velocity, ω, of the body via

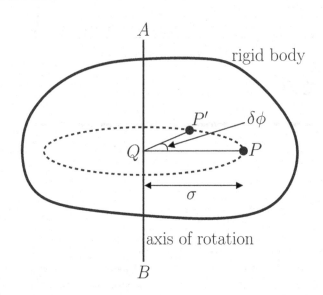

A

rigid body

P'

$\delta\phi$

Q

P

σ

axis of rotation

B

Figure 8.1 Rigid body rotation

where σ is the perpendicular distance from the axis of rotation to point P. (See Section 7.2.) Thus, in a rigidly rotating body, the rotation speed increases linearly with (perpendicular) distance from the axis of rotation.

It is helpful to introduce the *angular acceleration*, $\alpha(t)$, of a rigidly rotating body; this quantity is defined as the time derivative of the angular velocity. Thus,

$$\alpha \equiv \frac{d\omega}{dt} = \frac{d^2\phi}{dt^2}, \tag{8.3}$$

where ϕ is the angular coordinate of some arbitrarily chosen point reference within the body, measured with respect to the rotation axis. Note that angular velocities are conventionally measured in radians per second, whereas angular accelerations are measured in radians per second squared.

For a body rotating with constant angular velocity, ω, the angular acceleration is zero, and the rotation angle, ϕ, increases linearly with time, so that

$$\phi(t) = \phi_0 + \omega t, \tag{8.4}$$

where $\phi_0 = \phi(t = 0)$. Likewise, for a body rotating with constant angular acceleration, α, the angular velocity increases linearly with time, so that

$$\omega(t) = \omega_0 + \alpha t, \tag{8.5}$$

and the rotation angle satisfies

$$\phi(t) = \phi_0 + \omega_0 t + \frac{1}{2}\alpha t^2. \tag{8.6}$$

Here, $\omega_0 = \omega(t = 0)$. Note that there is a clear analogy between the previous two equations, and the equations of rectilinear motion at constant acceleration introduced in Section 2.6.

8.3   IS ROTATION A VECTOR?

Consider a rigid body that rotates through an angle ϕ about a given axis. It is tempting to try to define a rotation "vector", $\boldsymbol{\phi}$, that describes this motion. For example, suppose that $\boldsymbol{\phi}$ is defined as the "vector" whose magnitude is the angle of rotation, ϕ, and whose direction runs parallel to the axis of rotation. Unfortunately, this definition is ambiguous, because there are two possible directions that run parallel to the rotation axis. However, we can resolve this problem by adopting the following convention; the rotation "vector" runs parallel to the axis of rotation in the sense indicated by the thumb of the right-hand, when the fingers of this hand circulate around the axis in the direction of rotation. This convention is known as the *right-hand grip rule*.

The rotation "vector", $\boldsymbol{\phi}$, now has a well-defined magnitude and direction. But, is this quantity really a vector? This may seem like a strange question to ask, but it turns out that not all quantities that have well-defined magnitudes and directions are necessarily vectors. Let us review some properties of vectors. If **a** and **b** are two general vectors then it is certainly the case that

$$\mathbf{a} + \mathbf{b} = \mathbf{b} + \mathbf{a}. \tag{8.7}$$

(See Section 3.2.2.) In other words, the addition of vectors is necessarily commutative (i.e., it is independent of the order of addition). Is this true for "vector" rotations, as we have just defined them? Figure 8.2 shows the effect of applying two successive 90° rotations—one about the x-axis, and the other about the z-axis—to a six-sided die. In the left-hand case, the z-rotation is applied before the x-rotation, and vice versa in the right-hand case. It can be seen that the die ends up in two completely different states. Clearly, the z-rotation plus the x-rotation does not equal the x-rotation plus the z-rotation. This non-commutative algebra cannot be represented by vectors. We conclude that, although rotations have well-defined magnitudes and directions, rotations are not, in general, vector quantities.

There is a direct analogy between rotation and motion over the Earth's surface. After all, the motion of a pointer along the Earth's equator from longitude 0°W to longitude 90°W could just as well be achieved by keeping the pointer fixed and rotating the Earth through 90° about a north-south axis. The non-commutative nature of rotation "vectors" is a direct consequence of the non-planar (i.e., curved) nature of the Earth's surface. For instance, suppose we start off at (0° N, 0° W), which is just off the Atlantic coast of equatorial Africa, and rotate 90° northward and then 90° eastward. We end up at (0° N, 90° E), which is in the middle of the Indian Ocean. However, if we start at the same initial point, and rotate 90° eastward and then 90° northward, then we end up at the North Pole. Hence, large rotations over the Earth's surface do not commute. Let us now repeat this experiment on a far smaller scale. Suppose that we walk 10 m northward and then 10 m eastward. Next, suppose that, starting from the same initial position, we walk 10 m eastward and then 10 m northward. In this case, few people would dispute that the two end points are essentially identical. The crucial point is that, for sufficiently small displacements, the Earth's surface is approximately planar, and vector displacements on a plane surface commute with one another. This observation immediately suggests that rotation "vectors" that correspond to rotations through small angles must also commute with one another. In other words, although the quantity $\boldsymbol{\phi}$, defined previously, is not a true vector, the infinitesimal quantity $\delta\boldsymbol{\phi}$, which is defined in a similar manner, but corresponds to a rotation through an infinitesimal angle, $\delta\phi$, is a perfectly good vector.

We have just established that it is possible to define a true vector, $\delta\boldsymbol{\phi}$, that describes a rotation through a small angle, $\delta\phi$, about a fixed axis. But, how is this definition useful? Suppose that vector $\delta\boldsymbol{\phi}$ describes the small rotation that a given object executes in the

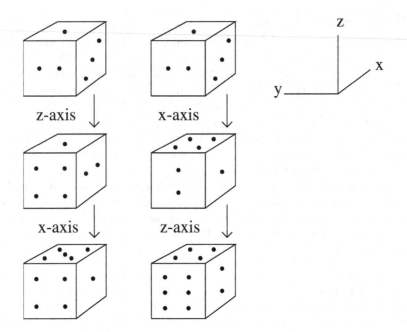

Figure 8.2 The addition of rotation is non-commutative. (Reproduced from Fitzpatrick 2008. Courtesy of Jones & Bartlett.)

infinitesimal time interval between t and $t + \delta t$. We can then define the quantity

$$\boldsymbol{\omega} \equiv \lim_{\delta t \to 0} \frac{\delta \boldsymbol{\phi}}{\delta t} = \frac{d\boldsymbol{\phi}}{dt}. \tag{8.8}$$

This quantity is clearly a true vector, because it is the ratio of a true vector and a scalar. Of course, $\boldsymbol{\omega}$ represents an *angular velocity vector*. The magnitude of this vector, ω, specifies the instantaneous angular velocity of the object, whereas the direction of the vector indicates the axis of rotation. The sense of rotation is given by the right-hand grip rule; if the thumb of the right-hand points along the direction of the vector then the fingers of the right-hand indicate the sense of rotation. We conclude that, although rotation can only be thought of as a vector quantity under certain very special circumstances, we can safely treat angular velocity as a vector quantity under all circumstances.

Suppose, for example, that a rigid body rotates at constant angular velocity $\boldsymbol{\omega}_1$. Let us now combine this motion with rotation about a different axis at the constant angular velocity $\boldsymbol{\omega}_2$. What is the subsequent motion of the body? Because we know that angular velocity is a vector, we can be certain that the combined motion simply corresponds to rotation about a third axis at constant angular velocity

$$\boldsymbol{\omega}_3 = \boldsymbol{\omega}_1 + \boldsymbol{\omega}_2, \tag{8.9}$$

where the sum is performed according to the standard rules of vector addition. [The previous result is subject to following important proviso. In order for Equation (8.9) to be valid, the rotation axes corresponding to $\boldsymbol{\omega}_1$ and $\boldsymbol{\omega}_2$ must cross at a certain point; the rotation axis corresponding to $\boldsymbol{\omega}_3$ then also passes through this point.] Moreover, a constant angular velocity

$$\boldsymbol{\omega} = \omega_x \, \mathbf{e}_x + \omega_y \, \mathbf{e}_y + \omega_z \, \mathbf{e}_z \tag{8.10}$$

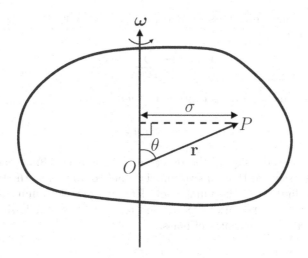

Figure 8.3 Rigid body rotation. The rotation velocity at point P is into the page.

can be thought of as representing rotation about the x-axis at angular velocity ω_x, combined with rotation about the y-axis at angular velocity ω_y, combined with rotation about the z-axis at angular velocity ω_z. (There is, again, a proviso; namely, that the rotation axis corresponding to ω must pass through the origin. Of course, we can always shift the origin to ensure that this is the case.) Clearly, the knowledge that angular velocity is vector quantity can be extremely useful.

Figure 8.3 shows a rigid body rotating with angular velocity ω. For the sake of simplicity, the axis of rotation, which runs parallel to ω, is assumed to pass through the origin, O, of our coordinate system. Point P, whose position vector is $\mathbf{r}$, represents a general point inside the body. What is the velocity of rotation, $\mathbf{v}$, at point P? The magnitude of this velocity is simply

$$v = \sigma \omega = \omega r \sin \theta, \qquad (8.11)$$

where σ is the perpendicular distance of point P from the axis of rotation, and θ is the angle subtended between the directions of ω and $\mathbf{r}$. The direction of the velocity is into the page. Another way of saying this is that the direction of the velocity is mutually perpendicular to the directions of ω and $\mathbf{r}$, in the sense indicated by the right-hand grip rule when ω is rotated onto $\mathbf{r}$ (through an angle less than $180°$). It follows that we can write

$$\mathbf{v} = \omega \times \mathbf{r}. \qquad (8.12)$$

(See Section 3.2.6.) Note, incidentally, that the direction of the angular velocity vector, ω, indicates the orientation of the axis of rotation. However, nothing actually moves in this direction. In fact, all of the motion is perpendicular to the direction of ω.

8.4 CENTER OF MASS

The *center of mass*—or center of gravity—of an extended object is defined in much the same manner as we earlier defined the center of mass of a set of mutually interacting point mass objects. See Section 6.3. To be more exact, the coordinates of the center of mass of an extended object are the mass-weighted averages of the coordinates of the elements that make up that object. Thus, if the object has net mass M, and is composed of N elements, such that the ith element has mass m_i and position vector $\mathbf{r}_i$, then the position vector of the center of mass is given by

$$\mathbf{r}_{cm} = \frac{1}{M} \sum_{i=1,N} m_i \, \mathbf{r}_i. \tag{8.13}$$

If the object under consideration is continuous then

$$m_i = \rho(\mathbf{r}_i) \, V_i, \tag{8.14}$$

where $\rho(\mathbf{r})$ is the mass density of the object and V_i is the volume occupied by the ith element. Here, it is assumed that this volume is small compared to the total volume of the object. Taking the limit that the number of elements goes to infinity and the volume of each element goes to zero, Equations (8.13) and (8.14) yield the following integral formula for the position vector of the center of mass:

$$\mathbf{r}_{cm} = \frac{1}{M} \int \rho \, \mathbf{r} \, dV. \tag{8.15}$$

Here, the integral is taken over the whole volume of the object, and $dV = dx \, dy \, dz$ is an element of that volume. Finally, for an object whose mass density is constant—which is the only type of object that we shall consider in this chapter—the previous expression reduces to

$$\mathbf{r}_{cm} = \frac{1}{V} \int \mathbf{r} \, dV, \tag{8.16}$$

where V is the volume of the object. According to Equation (8.16), the center of mass of a body of uniform density is located at the geometric center, or *centroid*, of that body.

For many solid objects, the location of the geometric center follows from symmetry. For instance, the geometric center of a cube is the point of intersection of the cube's long diagonals, drawn linking diagrammatically opposite corners. Likewise, the geometric center of a right cylinder is located on the symmetry axis, half-way up the cylinder.

8.4.1   Centroid of Regular Pyramid

As an illustration of the use of formula (8.16), let us calculate the geometric center of a regular square-sided pyramid. Figure 8.4 shows such a pyramid. Let a be the length of each side. It follows, from simple trigonometry, that the height of the pyramid is $h = a/\sqrt{2}$. Suppose that the base of the pyramid lies on the x-y plane, and the apex is aligned with the z-axis, as shown in the diagram. It follows, from symmetry, that the geometric center of the pyramid lies on the z-axis. It only remains to calculate the perpendicular distance, z_{cm}, between the geometric center and the base of the pyramid. This quantity is obtained from the z-component of Equation (8.16):

$$z_{cm} = \frac{\int\int\int z \, dx \, dy \, dz}{\int\int\int dx \, dy \, dz}, \tag{8.17}$$

where the integral is taken over the volume of the pyramid.

In the previous integral, the limits of integration for z are $z = 0$ to $z = h$, respectively (i.e., from the base to the apex of the pyramid). The corresponding limits of integration for x and y are $x, y = -a\,(1 - z/h)/2$ to $x, y = +a\,(1 - z/h)/2$, respectively (i.e., the limits are $x, y = \pm a/2$ at the base of the pyramid, and $x, y = \pm 0$ at the apex). Hence, Equation (8.17)

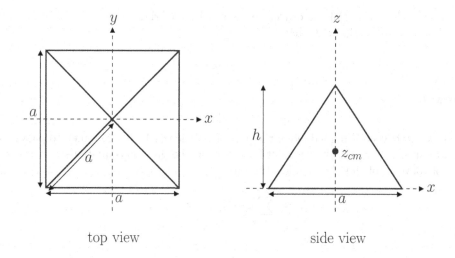

top view   side view

Figure 8.4 Locating the geometric center of a regular square-sided pyramid

can be written more explicitly as

$$z_{cm} = \frac{\int_0^h z\,dz \int_{-a\,(1-z/h)/2}^{+a\,(1-z/h)/2} dy \int_{-a\,(1-z/h)/2}^{+a\,(1-z/h)/2} dx}{\int_0^h dz \int_{-a\,(1-z/h)/2}^{+a\,(1-z/h)/2} dy \int_{-a\,(1-z/h)/2}^{+a\,(1-z/h)/2} dx}. \tag{8.18}$$

As indicated, it makes sense to perform the x- and y-integrals before the z-integrals, because the limits of integration for the x- and y-integrals are z-dependent. Performing the x-integrals, we obtain

$$z_{cm} = \frac{\int_0^h z\,dz \int_{-a\,(1-z/h)/2}^{+a\,(1-z/h)/2} a\,(1-z/h)\,dy}{\int_0^h dz \int_{-a\,(1-z/h)/2}^{+a\,(1-z/h)/2} a\,(1-z/h)\,dy}. \tag{8.19}$$

Performing the y-integrals, we get

$$z_{cm} = \frac{\int_0^h a^2\,z\,(1-z/h)^2\,dz}{\int_0^h a^2\,(1-z/h)^2\,dz}. \tag{8.20}$$

Finally, performing the z-integrals, we obtain

$$z_{cm} = \frac{a^2\left[z^2/2 - 2\,z^3/(3\,h) + z^4/(4\,h^2)\right]_0^h}{a^2\left[z - z^2/(h) + z^3/(3\,h)\right]_0^h} = \frac{a^2\,h^2/12}{a^2\,h/3} = \frac{h}{4}. \tag{8.21}$$

Thus, the geometric center of a regular square-sided pyramid is located on the symmetry axis, one quarter of the way from the base to the apex.

8.5 MOMENT OF INERTIA

Consider an extended object that is made up of N elements. Let the ith element possess mass m_i, position vector $\mathbf{r}_i$, and velocity $\mathbf{v}_i$. The total kinetic energy of the object is written

$$K = \sum_{i=1,N} \frac{1}{2}\,m_i\,v_i^2. \tag{8.22}$$

Suppose that the motion of the object consists merely of rigid rotation at angular velocity $\boldsymbol{\omega}$. It follows, from Section 8.3, that

$$\mathbf{v}_i = \boldsymbol{\omega} \times \mathbf{r}_i. \tag{8.23}$$

Let us write

$$\boldsymbol{\omega} = \omega \, \mathbf{k}, \tag{8.24}$$

where $\mathbf{k}$ is a unit vector aligned along the axis of rotation (which is assumed to pass through the origin of our coordinate system). It follows from the previous equations that the kinetic energy of rotation of the object takes the form

$$K = \sum_{i=1,N} \frac{1}{2} \, m_i \, |\mathbf{k} \times \mathbf{r}_i|^2 \, \omega^2, \tag{8.25}$$

or

$$K = \frac{1}{2} \, I \, \omega^2. \tag{8.26}$$

Here, the quantity I is termed the *moment of inertia* of the object, and is written

$$I = \sum_{i=1,N} m_i \, |\mathbf{k} \times \mathbf{r}_i|^2 = \sum_{i=1,N} m_i \, \sigma_i^2, \tag{8.27}$$

where $\sigma_i = |\mathbf{k} \times \mathbf{r}_i|$ is the perpendicular distance from the ith element to the axis of rotation. Note that for translational motion we usually write

$$K = \frac{1}{2} \, M \, v^2, \tag{8.28}$$

where M represents mass and v represents speed. A comparison of Equations (8.26) and (8.28) suggests that moment of inertia plays the same role in rotational motion that mass plays in translational motion.

For a continuous object, analogous arguments to those employed in Section 8.4 yield

$$I = \int \rho \, \sigma^2 \, dV, \tag{8.29}$$

where $\rho(\mathbf{r})$ is the mass density of the object, $\sigma = |\mathbf{k} \times \mathbf{r}|$ is the perpendicular distance from the axis of rotation, and dV is a volume element. Finally, for an object of constant density, the previous expression reduces to

$$I = M \, \frac{\int \sigma^2 \, dV}{\int dV}. \tag{8.30}$$

Here, M is the total mass of the object. Note that the integrals are taken over the whole volume of the object.

The moment of inertia of a uniform object depends not only on the size and shape of that object, but also on the location of the axis about which the object is rotating. In particular, the same object can have different moments of inertia when rotating about different axes.

Unfortunately, the evaluation of the moment of inertia of a given body about a given axis invariably involves the performance of a complicated volume integral. In fact, there is only one trivial moment of inertia calculation; namely, the moment of inertia of a thin circular ring about a symmetric axis that runs perpendicular to the plane of the ring. Suppose that

M is the mass of the ring and b is its radius. Each element of the ring shares a common perpendicular distance from the axis of rotation; namely, $\sigma = b$. Hence, Equation (8.30) reduces to

$$I = M\,b^2. \tag{8.31}$$

In general, moments of inertia are rather tedious to calculate. Fortunately, there exist two powerful theorems that enable us to relate the moment of inertia of a given body about a given axis to the moment of inertia of the same body about another axis.

8.5.1 Perpendicular Axis Theorem

The first of these theorems is called the *perpendicular axis theorem*, and only applies to uniform laminar (i.e., flat) objects. Consider a laminar object (i.e., a thin, planar object) of uniform density. Suppose, for the sake of simplicity, that the object lies in the x-y plane. The moment of inertia of the object about the z-axis is given by

$$I_z = M\,\frac{\int\int (x^2+y^2)\,dx\,dy}{\int\int dx\,dy}, \tag{8.32}$$

where we have suppressed the trivial z-integration, and the integral is taken over the extent of the object in the x-y plane. Incidentally, the previous expression follows from the observation that $\sigma^2 = x^2 + y^2$ when the axis of rotation is coincident with the z-axis. Likewise, the moments of inertia of the object about the x- and y-axes take the forms

$$I_x = M\,\frac{\int\int y^2\,dx\,dy}{\int\int dx\,dy}, \tag{8.33}$$

$$I_y = M\,\frac{\int\int x^2\,dx\,dy}{\int\int dx\,dy}, \tag{8.34}$$

respectively. Here, we have made use of the fact that $z = 0$ inside the object. It follows by inspection of the previous three equations that

$$I_z = I_x + I_y, \tag{8.35}$$

which is the essence of the perpendicular axis theorem.

Let us use the perpendicular axis theorem to find the moment of inertia of a thin ring about a symmetric axis that lies in the plane of the ring. Let the plane of the ring correspond to the x-y plane, and let the center of the ring correspond to the origin of the coordinate system. It is clear, from symmetry, that $I_x = I_y$. Now, we already know that $I_z = M\,b^2$, where M is the mass of the ring, and b is its radius. Hence, the perpendicular axis theorem tells us that

$$2\,I_x = I_z, \tag{8.36}$$

or

$$I_x = \frac{I_z}{2} = \frac{1}{2}\,M\,b^2. \tag{8.37}$$

Of course, $I_z > I_x$, because when the ring spins about the z-axis its elements are, on average, further from the axis of rotation than when it spins about the x-axis.

8.5.2 Parallel Axis Theorem

The second useful theorem regarding moments of inertia is called the *parallel axis theorem*. The parallel axis theorem, which is quite general, states that if I is the moment of inertia of a given body about an axis passing through the center of mass of that body then the moment of inertia, I', of the same body about a second axis that is parallel to the first is

$$I' = I + M d^2, \tag{8.38}$$

where M is the mass of the body and d is the perpendicular distance between the two axes.

In order to prove the parallel axis theorem, let us choose the origin of our coordinate system to coincide with the center of mass of the body in question. Furthermore, let us orientate the axes of our coordinate system such that the z-axis coincides with the first axis of rotation, whereas the second axis pierces the x-y plane at $x = d, y = 0$. From Equation (8.16), the fact that the center of mass is located at the origin implies that

$$\iiint x \, dx \, dy \, dz = \iiint y \, dx \, dy \, dz = \iiint z \, dx \, dy \, dz = 0, \tag{8.39}$$

where the integrals are taken over the volume of the body. From Equation (8.30), the expression for the first moment of inertia is

$$I = M \frac{\iiint (x^2 + y^2) \, dx \, dy \, dz}{\iiint dx \, dy \, dz}, \tag{8.40}$$

because $x^2 + y^2$ is the perpendicular distance of a general point (x, y, z) from the z-axis. Likewise, the expression for the second moment of inertia takes the form

$$I' = M \frac{\iiint [(x - d)^2 + y^2] \, dx \, dy \, dz}{\iiint dx \, dy \, dz}. \tag{8.41}$$

The previous equation can be expanded to give

$$I' = M \frac{\iiint [(x^2 + y^2) - 2 \, d \, x + d^2] \, dx \, dy \, dz}{\iiint dx \, dy \, dz}$$

$$= M \frac{\iiint (x^2 + y^2) \, dx \, dy \, dz}{\iiint dx \, dy \, dz} - 2 \, d \, M \frac{\iiint x \, dx \, dy \, dz}{\iiint dx \, dy \, dz}$$

$$+ d^2 \, M \frac{\iiint dx \, dy \, dz}{\iiint dx \, dy \, dz}. \tag{8.42}$$

It follows from Equations (8.39) and (8.40) that

$$I' = I + M d^2, \tag{8.43}$$

which proves the theorem.

Let us use the parallel axis theorem to calculate the moment of inertia, I', of a thin ring about an axis that runs perpendicular to the plane of the ring, and passes through the circumference of the ring. We know that the moment of inertia of a ring of mass M and radius b about an axis that runs perpendicular to the plane of the ring, and passes through the center of the ring—which coincides with the center of mass of the ring—is $I = M b^2$. Our new axis is parallel to this original axis, but shifted sideways by the perpendicular distance b. Hence, the parallel axis theorem tells us that

$$I' = I + M b^2 = 2 M b^2. \tag{8.44}$$

8.5.3   Moment of Inertia of a Circular Disk

As an illustration of the direct application of formula (8.30), let us calculate the moment of inertia of a thin uniform circular disk, of mass M and radius b, about an axis that passes through the center of the disk, and runs perpendicular to the plane of the disk. Let us choose our coordinate system such that the disk lies in the x-y plane, with its center at the origin. The axis of rotation is, therefore, coincident with the z-axis. Hence, formula (8.30) reduces to

$$I = M \frac{\iint (x^2 + y^2)\, dx\, dy}{\iint dx\, dy}, \tag{8.45}$$

where the integrals are taken over the area of the disk, and the redundant z-integration has been suppressed. Let us divide the disk up into thin annuli. Consider an annulus of radius $\sigma = \sqrt{x^2 + y^2}$ and radial thickness $d\sigma$. The area of this annulus is simply $2\pi\,\sigma\,d\sigma$. Hence, we can replace $dx\,dy$ in the previous integrals by $2\pi\,\sigma\,d\sigma$, so as to give

$$I = M \frac{\int_0^b 2\pi\,\sigma^3\,d\sigma}{\int_0^b 2\pi\,\sigma\,d\sigma}. \tag{8.46}$$

The previous expression yields

$$I = M \frac{[2\pi\,\sigma^4/4]_0^b}{[2\pi\,\sigma^2/2]_0^b} = \frac{1}{2}\,M\,b^2. \tag{8.47}$$

8.5.4   Standard Moments of Inertia

Similar calculations to the previous one yield the following standard results:

- The moment of inertia of a thin uniform rod of mass M and length l about an axis passing through the center of the rod and perpendicular to its length is

$$I = \frac{1}{12}\,M\,l^2.$$

- The moment of inertia of a thin uniform rod of mass M and length l about an axis passing through the end of the rod and perpendicular to its length is

$$I = \frac{1}{3}\,M\,l^2.$$

- The moment of inertia of a thin uniform rectangular sheet of mass M and dimensions a and b about a perpendicular axis passing through the center of the sheet is

$$I = \frac{1}{12}\,M\,(a^2 + b^2).$$

- The moment of inertia of a solid uniform cylinder of mass M and radius b about the cylindrical axis is

$$I = \frac{1}{2}\,M\,b^2.$$

- The moment of inertia of a thin uniform spherical shell of mass M and radius b about a diameter is

$$I = \frac{2}{3} M b^2.$$

- The moment of inertia of a solid uniform sphere of mass M and radius b about a diameter is

$$I = \frac{2}{5} M b^2.$$

8.6 TORQUE

We have now identified the rotational equivalent of velocity—namely, angular velocity—and the rotational equivalent of mass—namely, moment of inertia. But, what is the rotational equivalent of force?

Consider a bicycle wheel of radius b that is free to rotate around a perpendicular axis passing through its center. Suppose that we apply a force, $\mathbf{f}$, that is coplanar with the wheel, to a point P lying on its circumference. See Figure 8.5. What is the wheel's subsequent motion?

Let us choose the origin, O, of our coordinate system to coincide with the pivot point of the wheel; that is, the point of intersection between the wheel and the axis of rotation. Let $\mathbf{r}$ be the position vector of point P, and let θ be the angle subtended between the directions of $\mathbf{r}$ and $\mathbf{f}$. We can resolve $\mathbf{f}$ into two components; namely, a component, $f \cos\theta$, that acts radially, and a component, $f \sin\theta$, that acts tangentially. The radial component of $\mathbf{f}$ is canceled out by a reaction at the pivot, because the wheel is assumed to be mounted in such a manner that it can only rotate, and is prevented from displacing sideways. The tangential component of $\mathbf{f}$ causes the wheel to accelerate tangentially. Let v be the instantaneous rotation velocity of the wheel's circumference. Newton's second law of motion, applied to the tangential motion of the wheel, yields

$$M \dot{v} = f \sin\theta, \tag{8.48}$$

where M is the mass of the wheel (which is assumed to be concentrated in the wheel's rim).

Let us now convert the previous expression into a rotational equation of motion. If ω is the instantaneous angular velocity of the wheel then the relation between ω and v is simply

$$v = b\omega. \tag{8.49}$$

Because the wheel is basically a ring of radius b, rotating about a perpendicular symmetric axis, its moment of inertia is

$$I = M b^2. \tag{8.50}$$

Combining the previous three equations, we obtain

$$I \dot{\omega} = \tau, \tag{8.51}$$

where

$$\tau = f b \sin\theta. \tag{8.52}$$

Equation (8.51) is the *angular equation of motion* of the wheel. It relates the wheel's angular velocity, ω, and moment of inertia, I, to a quantity, τ, that is known as the *torque*.

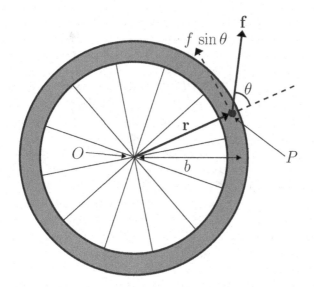

Figure 8.5 A rotating bicycle wheel

Clearly, if I is analogous to mass, and ω is analogous to velocity, then torque must be analogous to force. In other words, torque is the rotational equivalent of force.

It is clear, from Equation (8.52), that a torque is the product of the magnitude of the applied force, f, and some distance $l = b \sin \theta$. The physical interpretation of l is illustrated in Figure 8.6. Here, O is the axis of rotation, and the force, **f**, is applied at point P. It can be seen that l is the perpendicular distance of the line of action of the force from the axis of rotation. We usually refer to this distance as the length of the *lever arm*. Note that the *line of action* of a force is a straight-line that passes through the point of action of the force (i.e., the point at which the force acts) and is parallel to the force.

In summary, a torque measures the propensity of a given force to cause the object upon which it acts to twist about a certain axis. The torque, τ, is simply the product of the magnitude of the applied force, f, and the length of the lever arm, l:

$$\tau = f\,l. \tag{8.53}$$

Of course, this definition makes a lot of sense. We all know that it is far easier to turn a rusty bolt using a long, rather than a short, wrench. Assuming that we exert the same force on the end of each wrench, the torque that we apply to the bolt is larger in the former case, because the perpendicular distance between the line of action of the force and the bolt (i.e., the length of the wrench) is greater.

Because force is a vector quantity, it stands to reason that torque must also be a vector quantity. It follows that Equation (8.53) defines the magnitude, τ, of some torque vector, $\boldsymbol{\tau}$. But, what is the direction of this vector? By convention, if a torque is such as to cause the object upon which it acts to twist about a certain axis then the direction of that torque runs along the direction of the axis in the sense given by the right-hand grip rule. In other words, if the fingers of the right-hand circulate around the axis of rotation in the sense in which the torque twists the object then the thumb of the right-hand points along the axis in the direction of the torque. It follows that we can rewrite our rotational equation of motion, Equation (8.51), in vector form:

$$I \frac{d\boldsymbol{\omega}}{dt} = I\,\boldsymbol{\alpha} = \boldsymbol{\tau}, \tag{8.54}$$

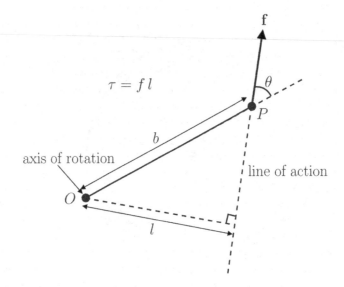

$$\tau = f\,l$$

Figure 8.6 Definition of the length of the lever arm, l

where $\alpha = d\omega/dt$ is the vector angular acceleration. Note that the direction of α indicates the direction of the rotation axis about which the object accelerates (in the sense given by the right-hand grip rule), whereas the direction of τ indicates the direction of the rotation axis about which the torque attempts to twist the object (in the sense given by the right-hand grip rule). Of course, these two rotation axes are identical.

Although Equation (8.54) was derived for the special case of a torque applied to a ring rotating about a perpendicular symmetric axis, it is, nevertheless, completely general.

It is important to appreciate that the directions that we ascribe to angular velocities, angular accelerations, and torques are merely conventions. There is actually no physical motion in the direction of the angular velocity vector; in fact, all of the motion is in the plane perpendicular to this vector. Likewise, there is no physical acceleration in the direction of the angular acceleration vector; again, all of the acceleration is in the plane perpendicular to this vector. Finally, no physical forces act in the direction of the torque vector; in fact, all of the forces act in the plane perpendicular to this vector.

Consider a rigid body that is free to pivot in any direction about some fixed point, O. Suppose that a force, $\mathbf{f}$, is applied to the body at some point, P, whose position vector relative to O is $\mathbf{r}$. See Figure 8.7. Let θ be the angle subtended between the directions of $\mathbf{r}$ and $\mathbf{f}$. What is the vector torque, τ, acting on the object about an axis passing through the pivot point? The magnitude of this torque is simply

$$\tau = r\,f\,\sin\theta. \qquad (8.55)$$

In Figure 8.7, the conventional direction of the torque is out of the page. (Because the force is clearly trying to twist the object in a counter-clockwise direction.) Another way of saying this is that the direction of the torque is mutually perpendicular to both $\mathbf{r}$ and $\mathbf{f}$, in the sense given by the right-hand grip rule when vector $\mathbf{r}$ is rotated onto vector $\mathbf{f}$ (through an angle less than 180° degrees). It follows that we can write

$$\tau = \mathbf{r} \times \mathbf{f}. \qquad (8.56)$$

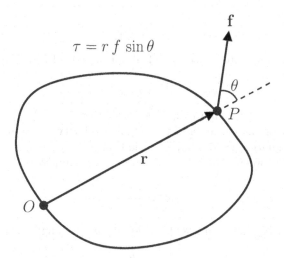

$$\tau = r\,f\,\sin\theta$$

Figure 8.7 Torque about a fixed point. The direction of the torque is out of the page.

In other words, the torque exerted by a force acting on a rigid body that pivots about some fixed point is the vector product of the displacement of the point of application of the force from the pivot point with the force itself. Equation (8.56) specifies both the magnitude of the torque, and the axis of rotation about which the torque twists the body upon which it acts. This axis runs parallel to the direction of $\boldsymbol{\tau}$, and passes through the pivot point.

8.7 POWER AND WORK

Consider a mass, m, attached to the end of a light rod of length l whose other end is attached to a fixed pivot. Suppose that the pivot is such that the rod is free to rotate in any direction. Suppose, further, that a force, $\mathbf{f}$, is applied to the mass, whose instantaneous angular velocity about an axis of rotation passing through the pivot is $\boldsymbol{\omega}$.

Let $\mathbf{v}$ be the instantaneous velocity of the mass. We know that the rate at which the force, $\mathbf{f}$, performs work on the mass, otherwise known as the power, is given by

$$P = \mathbf{f} \cdot \mathbf{v}. \tag{8.57}$$

(See Section 5.8.) However, we also know that (see Section 8.3)

$$\mathbf{v} = \boldsymbol{\omega} \times \mathbf{r}, \tag{8.58}$$

where $\mathbf{r}$ is the vector displacement of the mass from the pivot. Hence, we can write

$$P = \boldsymbol{\omega} \times \mathbf{r} \cdot \mathbf{f}. \tag{8.59}$$

(Note that $\mathbf{a} \cdot \mathbf{b} = \mathbf{b} \cdot \mathbf{a}$.)

Now, for any three vectors, $\mathbf{a}$, $\mathbf{b}$, and $\mathbf{c}$, we can write

$$\mathbf{a} \times \mathbf{b} \cdot \mathbf{c} = \mathbf{a} \cdot \mathbf{b} \times \mathbf{c}. \tag{8.60}$$

This theorem is easily proved by expanding the vector and scalar products in component form, using the definitions (3.24) and (3.33). It follows that Equation (8.59) can be rewritten

$$P = \boldsymbol{\omega} \cdot \mathbf{r} \times \mathbf{f}. \tag{8.61}$$

However,

$$\boldsymbol{\tau} = \mathbf{r} \times \mathbf{f}, \tag{8.62}$$

where $\boldsymbol{\tau}$ is the torque associated with force $\mathbf{f}$ about an axis of rotation passing through the pivot. Hence, we obtain

$$P = \boldsymbol{\tau} \cdot \boldsymbol{\omega}. \tag{8.63}$$

In other words, the rate at which a torque performs work on the object upon which it acts is the scalar product of the torque and the angular velocity of the object. Note the great similarity between Equation (8.57) and Equation (8.63).

The relationship between work, W, and power, P, is simply

$$P = \frac{dW}{dt}. \tag{8.64}$$

Likewise, the relationship between angular velocity, $\boldsymbol{\omega}$, and angle of rotation, $\boldsymbol{\phi}$, is

$$\boldsymbol{\omega} = \frac{d\boldsymbol{\phi}}{dt}. \tag{8.65}$$

It follows that Equation (8.63) can be rewritten

$$dW = \boldsymbol{\tau} \cdot d\boldsymbol{\phi}. \tag{8.66}$$

Integration yields

$$W = \int \boldsymbol{\tau} \cdot d\boldsymbol{\phi}. \tag{8.67}$$

Note that this is a good definition, because it only involves an infinitesimal rotation vector, $d\boldsymbol{\phi}$. (Recall, from Section 8.3, that it is impossible to define a finite rotation vector.) For the case of translational motion, the analogous expression to the previous one is

$$W = \int \mathbf{f} \cdot d\mathbf{r}. \tag{8.68}$$

Here, $\mathbf{f}$ is the force, and $d\mathbf{r}$ is an element of displacement of the body upon which the force acts.

Although Equations (8.63) and (8.67) were derived for the special case of the rotation of a mass attached to the end of a light rod, they are, nevertheless, completely general.

Consider, finally, the special case in which the torque is aligned with the angular velocity, and both are constant in time. In this case, the rate at which the torque performs work is simply

$$P = \tau \omega. \tag{8.69}$$

Likewise, the net work performed by the torque in twisting the body upon which it acts through an angle $\Delta\phi$ is just

$$W = \tau \, \Delta\phi. \tag{8.70}$$

8.8 TRANSLATIONAL MOTION VERSUS ROTATIONAL MOTION

It should be clear, by now, that there is a strong analogy between rotational motion and standard translational motion. Indeed, each physical concept used to analyze rotational motion has its translational concomitant. Likewise, every law of physics governing rotational motion has a translational equivalent. The analogies between rotational and translational motion are summarized in Table 8.1.

Table 8.1   The Analogies Between Translational and Rotational Motion

Translational motion		Rotational motion	
Displacement	$d\mathbf{r}$	Angular displacement	$d\phi$
Velocity	$\mathbf{v} = d\mathbf{r}/dt$	Angular velocity	$\boldsymbol{\omega} = d\phi/dt$
Acceleration	$\mathbf{a} = d\mathbf{v}/dt$	Angular acceleration	$\boldsymbol{\alpha} = d\omega/dt$
Mass	M	Moment of inertia	$I = \int \rho\,|\mathbf{k} \times \mathbf{r}|^2\,dV$
Force	$\mathbf{f} = M\mathbf{a}$	Torque	$\boldsymbol{\tau} \equiv \mathbf{r} \times \mathbf{f} = I\boldsymbol{\alpha}$
Work	$W = \int \mathbf{f} \cdot d\mathbf{r}$	Work	$W = \int \boldsymbol{\tau} \cdot d\phi$
Power	$P = \mathbf{f} \cdot \mathbf{v}$	Power	$P = \boldsymbol{\tau} \cdot \boldsymbol{\omega}$
Kinetic energy	$K = M v^2/2$	Kinetic energy	$K = I\omega^2/2$

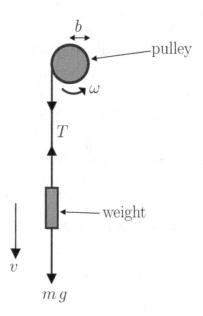

Figure 8.8   An unwinding pulley

8.9   UNWINDING PULLEY

Suppose that a weight of mass m is suspended via a light inextensible cable that is wound around a pulley of mass M and radius b. See Figure 8.8. Let us treat the pulley as a uniform disk, and let us assume that the cable does not slip with respect to the pulley. What is the downward acceleration of the weight, and the tension in the cable?

Let v be the instantaneous downward velocity of the weight, ω the instantaneous angular velocity of the pulley, and T the tension in the cable. Applying Newton's second law to the vertical motion of the weight, we obtain

$$m\dot{v} = mg - T. \tag{8.71}$$

The angular equation of motion of the pulley is written

$$I\dot{\omega} = \tau, \tag{8.72}$$

where I is its moment of inertia, and τ is the torque acting on the pulley. Now, the only force acting on the pulley (whose line of action does not pass through the pulley's axis of rotation) is the tension in the cable. The torque associated with this force is the product of the tension, T, and the perpendicular distance from the line of action of this force to the rotation axis, which is equal to the radius, b, of the pulley. Hence,

$$\tau = T\,b. \tag{8.73}$$

If the cable does not slip with respect to the pulley then its downward velocity, v, must match the tangential velocity of the outer surface of the pulley, $b\omega$. Thus,

$$v = b\,\omega. \tag{8.74}$$

It follows that

$$\dot{v} = b\,\dot{\omega}. \tag{8.75}$$

Equations (8.72), (8.73), and (8.75) can be combined to give

$$T = \frac{I}{b^2}\,\dot{v}. \tag{8.76}$$

It follows from Equations (8.71) and (8.76) that

$$\dot{v} = \frac{g}{1 + I/m\,b^2}, \tag{8.77}$$

$$T = \frac{m\,g}{1 + m\,b^2/I}. \tag{8.78}$$

Now, the moment of inertia of the pulley is $I = (1/2)\,M\,b^2$. Hence, the previous expressions reduce to

$$\dot{v} = \frac{g}{1 + M/2\,m}, \tag{8.79}$$

$$T = \frac{m\,g}{1 + 2\,m/M}. \tag{8.80}$$

Note that the downward acceleration of the weight is less than the acceleration due to gravity, and the tension in the cable is non-zero, because the rotational inertia of the pulley impedes the motion of the system.

8.10 PHYSICS OF BASEBALL BATS

Baseball players know from experience that there is a "sweet spot" on a baseball bat, about 17 cm from the end of the barrel, where the shock of impact with the ball, as felt by the hands, is minimized. In fact, if the ball strikes the bat exactly on the sweet spot then the hitter is almost unaware of the collision. Conversely, if the ball strikes the bat well away from the sweet spot then the impact is felt as a painful jarring of the hands.

The existence of a sweet spot on a baseball bat is a consequence of rotational dynamics. Let us analyze this problem. Consider the schematic baseball bat shown in Figure 8.9. Let M be the mass of the bat and let l be its length. Suppose that the bat pivots about a fixed point located at one of its ends. Let the center of mass of the bat be located a distance b from the pivot point. Finally, suppose that the ball strikes the bat a distance h from the pivot point.

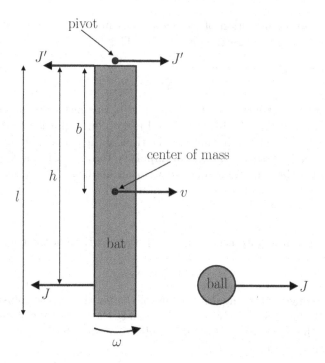

Figure 8.9 A schematic baseball bat

The collision between the bat and the ball can be modeled as equal and opposite impulses, J, applied to each object at the time of the collision. See Section 6.5. At the same time, equal and opposite impulses, J', are applied to the pivot and the bat, as shown in Figure 8.9. If the pivot actually corresponds to a hitter's hands then the latter impulse gives rise to the painful jarring sensation felt when the ball is not struck properly.

We saw earlier that in a general multi-component system, which includes an extended body such as a baseball bat, the motion of the center of mass takes a particularly simple form. See Section 6.3. To be more exact, the motion of the center of mass is equivalent to that of the point particle obtained by concentrating the whole mass of the system at the center of mass, and then allowing all of the external forces acting on the system to act upon that mass. Let us use this idea to analyze the effect of the collision with the ball on the motion of the bat's center of mass. The center of mass of the bat acts like a point particle of mass M that is subject to the two impulses, J and J' (which are applied simultaneously). If v is the instantaneous velocity of the center of mass then the change in momentum of this point due to the action of the two impulses is simply

$$M\,\Delta v = -J - J'. \tag{8.81}$$

(See Section 6.5.) The minus signs on the right-hand side of the previous equation follow from the fact that the impulses are oppositely directed to v in Figure 8.9.

Note that in order to specify the instantaneous state of an extended body, we must do more than just specify the location of the body's center of mass. Indeed, because the body can rotate about its center of mass, we must also specify its orientation in space. Thus, in order to follow the motion of an extended body, we must not only follow the translational motion of its center of mass, but also the body's rotational motion about this point (or any other convenient reference point located within the body).

Consider the rotational motion of the bat shown in Figure 8.9 about a perpendicular (to the bat) axis passing through the pivot point. This motion satisfies

$$I \frac{d\omega}{dt} = \tau, \tag{8.82}$$

where I is the moment of inertia of the bat, ω is its instantaneous angular velocity, and τ is the applied torque. The bat is actually subject to an impulsive torque (i.e., a torque that only lasts for a short period in time) at the time of the collision with the ball. Defining the angular impulse, K, associated with an impulsive torque, τ, in much the same manner as we earlier defined the impulse associated with an impulsive force (see Section 6.5), we obtain

$$K = \int^t \tau \, dt. \tag{8.83}$$

It follows that we can integrate Equation (8.82) over the time of the collision to find

$$I \, \Delta\omega = K, \tag{8.84}$$

where $\Delta\omega$ is the change in the angular velocity of the bat due to the collision with the ball.

The torque associated with a given force is equal to the magnitude of the force multiplied by the length of the lever arm. Thus, it stands to reason that the angular impulse, K, associated with an impulse, J, is simply

$$K = J \, x, \tag{8.85}$$

where x is the perpendicular distance from the line of action of the impulse to the axis of rotation. Hence, the angular impulses associated with the two impulses, J and J', to which the bat is subject when it collides with the ball, are $J \, h$ and 0, respectively. The latter angular impulse is zero because the point of application of the associated impulse coincides with the pivot point, and so the length of the lever arm is zero. It follows that Equation (8.84) can be written

$$I \, \Delta\omega = -J \, h. \tag{8.86}$$

The minus sign comes from the fact that the impulse J is oppositely directed to the angular velocity in Figure 8.9.

The relationship between the instantaneous velocity of the bat's center of mass and the bat's instantaneous angular velocity is simply

$$v = b\omega. \tag{8.87}$$

Hence, Equation (8.81) can be rewritten

$$M b \, \Delta\omega = -J - J'. \tag{8.88}$$

Equations (8.86) and (8.88) can be combined to yield

$$J' = -\left(1 - \frac{M b h}{I}\right) J. \tag{8.89}$$

The previous expression specifies the magnitude of the impulse, J', applied to the hitter's hands terms of the magnitude of the impulse, J, applied to the ball.

Let us crudely model the bat as a uniform rod of length l and mass M. It follows, by symmetry, that the center of mass of the bat lies at its midpoint; that is,

$$b = \frac{l}{2}. \tag{8.90}$$

Moreover, the moment of inertia of the bat about a perpendicular axis passing through one of its ends is

$$I = \frac{1}{3} M l^2. \tag{8.91}$$

(See Section 8.5.4.) Combining the previous three equations, we obtain

$$J' = -\left(1 - \frac{3h}{2l}\right) J = -\left(1 - \frac{h}{h_0}\right) J, \tag{8.92}$$

where

$$h_0 = \frac{2}{3} l. \tag{8.93}$$

Clearly, if $h = h_0$ then, no matter how hard the ball is hit (i.e., no matter how large we make J), zero impulse is applied to the hitter's hands. We conclude that the sweet spot—or, in scientific terms, the *center of percussion*—of a uniform baseball bat lies two-thirds of the way down the bat from the hitter's end. If we adopt a more realistic model of a baseball bat, in which the bat is tapered such that the majority of its weight is located at its hitting end, then we can easily demonstrate that the center of percussion is shifted further away from the hitter (i.e., it is more that two-thirds of the way along the bat).

8.11   COMBINED TRANSLATIONAL AND ROTATIONAL MOTION

In Section 4.11.1, we analyzed the motion of a block sliding down a frictionless incline. We found that the block accelerates down the slope with uniform acceleration $g \sin \theta$, where θ is the angle subtended by the incline with the horizontal. In this case, all of the potential energy lost by the block, as it slides down the slope, is converted into translational kinetic energy. (See Chapter 5.) In particular, no energy is dissipated.

It is not possible for a block to slide over a frictional surface without dissipating energy. However, we know from experience that a round object can roll over such a surface with hardly any dissipation. For instance, it is far easier to drag a heavy suitcase across the concourse of an airport if the suitcase has wheels on the bottom. Let us investigate the physics of round objects rolling over rough surfaces, and, in particular, rolling down rough inclines.

Consider a uniform cylinder of radius b rolling over a horizontal, frictional surface. See Figure 8.10. Let v be the translational velocity of the cylinder's center of mass and let ω be the angular velocity of the cylinder about an axis running along its length, and passing through its center of mass. Consider the point of contact between the cylinder and the surface. The velocity v' of this point is made up of two components; the translational velocity v, which is common to all elements of the cylinder, and the tangential velocity $v_t = -b\omega$, due to the cylinder's rotational motion. Thus,

$$v' = v + v_t = v - b\omega. \tag{8.94}$$

Suppose that the cylinder rolls without slipping. In other words, suppose that there is no frictional energy dissipation as the cylinder moves over the surface. This is only possible if

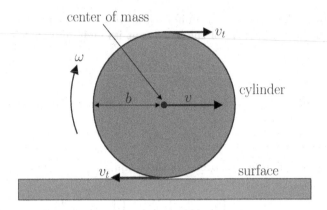

Figure 8.10 A cylinder rolling over a rough surface

there is zero net motion between the surface and the bottom of the cylinder, which implies $v' = 0$, or

$$v = b\,\omega. \tag{8.95}$$

It follows that if a cylinder, or any other round object, rolls across a rough surface without slipping—that is, without dissipating energy—then the cylinder's translational and rotational velocities are not independent, but satisfy a particular relationship. (See the previous equation.) Of course, if the cylinder slips as it rolls across the surface then this relationship no longer holds.

8.11.1   Cylinder Rolling Down a Rough Incline

Consider what happens when the cylinder shown in Figure 8.10 rolls, without slipping, down a rough slope whose angle of inclination, with respect to the horizontal, is θ. If the cylinder starts from rest, and rolls down the slope a vertical distance h, then its gravitational potential energy decreases by $-\Delta U = M\,g\,h$, where M is the mass of the cylinder. This decrease in potential energy must be offset by a corresponding increase in kinetic energy. (Recall that when a cylinder rolls without slipping there is no frictional energy loss.) However, a rolling cylinder can possesses two different types of kinetic energy. First, it can have translational kinetic energy, $K_t = (1/2)\,M\,v^2$, where v is the cylinder's translational velocity. Second, it can have rotational kinetic energy, $K_r = (1/2)\,I\,\omega^2$, where ω is the cylinder's angular velocity and I is its moment of inertia. Hence, energy conservation yields

$$M\,g\,h = \frac{1}{2}\,M\,v^2 + \frac{1}{2}\,I\,\omega^2. \tag{8.96}$$

When the cylinder rolls without slipping, its translational and rotational velocities are related via Equation (8.95). It follows from Equation (8.96) that

$$v^2 = \frac{2\,g\,h}{1 + I/M\,b^2}. \tag{8.97}$$

Making use of the fact that the moment of inertia of a uniform cylinder about its axis of symmetry is $I = (1/2)\,M\,b^2$, we can write the previous equation more explicitly as

$$v^2 = \frac{4}{3}\,g\,h. \tag{8.98}$$

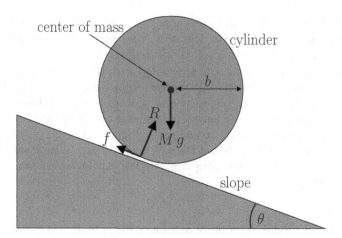

Figure 8.11 A cylinder rolling down a rough incline

If the same cylinder were to slide down a frictionless slope, such that it fell from rest through a vertical distance h, then its final translational velocity would satisfy

$$v^2 = 2\,g\,h. \tag{8.99}$$

A comparison of Equations (8.98) and (8.99) reveals that if a uniform cylinder rolls down an incline without slipping then its final translational velocity is less than that obtained when the cylinder slides down the same incline without friction. The reason for this is that, in the former case, some of the potential energy released as the cylinder falls is converted into rotational kinetic energy, whereas, in the latter case, all of the released potential energy is converted into translational kinetic energy. Note that, in both cases, the cylinder's total kinetic energy at the bottom of the incline is equal to the released potential energy.

Let us examine the equations of motion of a cylinder, of mass M and radius b, rolling down a rough slope without slipping. As shown in Figure 8.11, there are three forces acting on the cylinder. First, we have the cylinder's weight, $M\,g$, which acts vertically downward. Second, we have the reaction, R, of the slope, which acts normally outward from the surface of the slope. Third, we have the frictional force, f, which acts up the slope, parallel to its surface.

As we have already discussed, we can most easily describe the translational motion of an extended body by following the motion of its center of mass. This motion is equivalent to that of a point particle, whose mass equals that of the body, which is subject to the same external forces as those that act on the body. Thus, applying the three forces, $M\,g$, R, and f, to the cylinder's center of mass, and resolving in the direction normal to the surface of the slope, we obtain

$$R = M\,g\,\cos\theta. \tag{8.100}$$

Furthermore, Newton's second law, applied to the motion of the center of mass parallel to the slope, yields

$$M\,\dot{v} = M\,g\,\sin\theta - f, \tag{8.101}$$

where $\dot{v}$ is the cylinder's translational acceleration down the slope.

Let us, now, examine the cylinder's rotational equation of motion. First, we must evaluate the torques associated with the three forces acting on the cylinder. Recall, that the torque associated with a given force is the product of the magnitude of that force and the length of the level arm; that is, the perpendicular distance between the line of action of the force and the axis of rotation. By definition, the weight of an extended object acts at its center of mass. However, in this case, the axis of rotation passes through the center of mass. Hence, the length of the lever arm associated with the weight, $M g$, is zero. It follows that the associated torque is also zero. It is clear, from Figure 8.11, that the line of action of the reaction force, R, passes through the center of mass of the cylinder, which coincides with the axis of rotation. Thus, the length of the lever arm associated with R is zero, and so is the associated torque. Finally, according to Figure 8.11, the perpendicular distance between the line of action of the friction force, f, and the axis of rotation is just the radius of the cylinder, b, so the associated torque is $f b$. We conclude that the net torque acting on the cylinder is simply

$$\tau = f\,b. \tag{8.102}$$

It follows that the rotational equation of motion of the cylinder takes the form,

$$I\,\dot{\omega} = \tau = f\,b, \tag{8.103}$$

where I is its moment of inertia and $\dot{\omega}$ is its rotational acceleration.

If the cylinder rolls, without slipping, such that the constraint (8.95) is satisfied at all times, then the time derivative of this constraint implies the following relationship between the cylinder's translational and rotational accelerations:

$$\dot{v} = b\,\dot{\omega}. \tag{8.104}$$

It follows from Equations (8.101), (8.103), and (8.104) that

$$\dot{v} = \frac{g\,\sin\theta}{1 + I/M\,b^2}, \tag{8.105}$$

$$f = \frac{M\,g\,\sin\theta}{1 + M\,b^2/I}. \tag{8.106}$$

Because the moment of inertia of the cylinder is actually $I = (1/2)\,M\,b^2$, the previous expressions simplify to give

$$\dot{v} = \frac{2}{3}\,g\,\sin\theta, \tag{8.107}$$

and

$$f = \frac{1}{3}\,M\,g\,\sin\theta. \tag{8.108}$$

Note that the acceleration of a uniform cylinder as it rolls down a slope, without slipping, is only two-thirds of the value obtained when the cylinder slides down the same slope without friction. It is clear from Equation (8.101) that, in the former case, the acceleration of the cylinder down the slope is retarded by friction. Note, however, that the frictional force merely acts to convert translational kinetic energy into rotational kinetic energy, and does not dissipate energy.

In order for the slope to exert the frictional force specified in Equation (8.108), without any slippage between the slope and cylinder, this force must be less than the maximum allowable static frictional force, $\mu\,R = \mu\,M\,g\,\cos\theta$, where μ is the coefficient of static friction. (See Section 4.10.) In other words, the condition for the cylinder to roll down the slope without slipping is $f < \mu\,R$, or

$$\tan\theta < 3\,\mu. \tag{8.109}$$

This condition is easily satisfied for gentle slopes, but may well be violated for extremely steep slopes (depending on the size of μ). Of course, the previous condition is always violated for frictionless slopes, for which $\mu = 0$.

Suppose, finally, that we place two cylinders, side by side and at rest, at the top of a frictional slope of inclination θ. Let the two cylinders possess the same mass, M, and the same radius, b. However, suppose that the first cylinder is uniform, whereas the second is a hollow shell. Which cylinder reaches the bottom of the slope first, assuming that they are both released simultaneously, and both roll without slipping? The acceleration of each cylinder down the slope is given by Equation (8.105). For the case of the solid cylinder, the moment of inertia is $I = (1/2) M b^2$, and so

$$\dot{v}_{\text{solid}} = \frac{2}{3} g \sin \theta. \tag{8.110}$$

For the case of the hollow cylinder, the moment of inertia is $I = M b^2$ (i.e., the same as that of a ring with a similar mass, radius, and axis of rotation), and so

$$\dot{v}_{\text{hollow}} = \frac{1}{2} g \sin \theta. \tag{8.111}$$

It is clear that the solid cylinder reaches the bottom of the slope before the hollow one (because it possesses the greater acceleration). Note that the accelerations of the two cylinders are independent of their sizes or masses. This suggests that a solid cylinder will always roll down a frictional incline faster than a hollow one, irrespective of their relative dimensions (assuming that they both roll without slipping). In fact, Equation (8.105) suggests that if two different objects roll (without slipping) down the same slope then the most compact object—that is, the object with the smallest $I/M b^2$ ratio—always wins the race.

8.12 EXERCISES

8.1 A tire placed on a balancing machine in a service station starts from rest and turns through 5.3 revolutions in 2.3 s before reaching its final angular speed.

(a) What is the angular acceleration of the tire (assuming that this quantity remains constant)? [Ans: $12.59 \, \text{rad/s}^2$.]

(b) What is the final angular speed of the tire? [Ans: $3.142 \, \text{rad/s}$.]

8.2 The net work done in accelerating a wheel from rest to an angular speed of 30 rev/min is $W = 5500 \, \text{J}$. What is the moment of inertia of the wheel? [Ans: $1114.6 \, \text{kg m}^2$.]

8.3 A rod of mass $M = 3 \, \text{kg}$ and length $L = 1.2 \, \text{m}$ pivots about a fixed axis, perpendicular to its length, that passes through one of its ends.

(a) What is the moment of inertia of the rod? [Ans: $1.44 \, \text{kg m}^2$.]

(b) Given that the rod's instantaneous angular velocity is 60 deg/s, what is its rotational kinetic energy? [Ans: 0.789 J.]

8.4 A uniform rod of mass $m = 5.3 \, \text{kg}$ and length $l = 1.3 \, \text{m}$ rotates about a fixed, frictionless, horizontal axle, perpendicular to its length, located at one of its ends. The rod is released from rest at an angle $\theta = 35°$ beneath the horizontal. What is the angular acceleration of the rod immediately after it is released? [Ans: $9.26 \, \text{rad/s}^2$.]

8.5 A car engine develops a torque of $\tau = 500 \, \text{N m}$ and rotates at 3000 rev/min. What horsepower does the engine generate? (1 hp = 746 W). [Ans: 210.5 hp.]

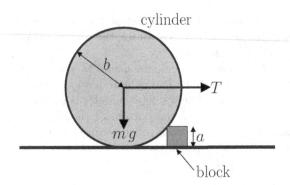

cylinder

Figure 8.12   Figure for Exercise 8.8

8.6 A uniform elliptical laminar of mass M lies in the x-y plane. The boundary of the laminar is specified by $x^2/a^2 + y^2/b^2 = 1$.

   (a) What is the moment of inertia of the laminar about the x-axis? [Ans: $(1/4)\, M\, b^2$.]

   (b) What is the moment of inertia of the laminar about the y-axis? [Ans: $(1/4)\, M\, a^2$.]

   (c) What is the moment of inertia of the laminar about the z-axis? [Ans: $(1/4)\, M\, (a^2 + b^2)$.]

8.7 The boundary of a uniform ellipsoid of mass M is specified by $x^2/a^2 + y^2/b^2 + z^2/c^2 = 1$.

   (a) What is the moment of inertia of the ellipsoid about the x-axis? [Ans: $(1/5)\, M\, (b^2 + c^2)$.]

   (b) What is the moment of inertia of the ellipsoid about the y-axis? [Ans: $(1/5)\, M\, (a^2 + c^2)$.]

   (c) What is the moment of inertia of the ellipsoid about the z-axis? [Ans: $(1/5)\, M\, (a^2 + b^2)$.]

8.8 What is the minimum horizontal force, T, required to make the cylinder in Figure 8.12 roll over the block? Let b be the radius of the cylinder, $a < b$ the height of the block, and m the mass of the cylinder. [Ans: $m\, g\, \sqrt{b^2 - (b-a)^2}/(b-a)$.]

8.9 Consider the table-pulley system shown in Figure 4.14. Suppose that the pulley has a radius b and a moment of inertia I. Suppose, further, that the cable does not slip over the surface of the pulley.

   (a) What is the acceleration of the system? [Ans: $m_2\, g/(m_1 + m_2 + I/b^2)$.]

   (b) What is the tension in the horizontal section of the cable? [Ans: $m_1\, m_2\, g/(m_1 + m_2 + I/b^2)$.]

   (c) What is the tension in the vertical section of the cable? [Ans: $(m_1 + I/b^2)\, m_2\, g/(m_1 + m_2 + I/b^2)$.]

8.10 A light cable is wrapped around a solid cylinder of mass M and radius b. The free end of the cable is held stationary, and the cylinder is released from rest. The cable unwinds from the cylinder, but does not slip, as the cylinder descends.

(a) What is the downward acceleration of the cylinder. [Ans: $(2/3)\,g$.]

(b) What is the tension in the cable? [Ans: $(1/3)\,M\,g$.]

8.11 Consider the Atwood machine shown in Figure 4.15. Suppose that the pulley has a radius b and a moment of inertia I. Suppose, further, that the cable does not slip over the surface of the pulley. What is the downward acceleration of the mass m_1? [Ans: $(m_1 - m_2)\,g/(m_1 + m_2 + I/b^2)$.]

8.12 A uniform cylinder of mass M and radius b has a light, thin cable wound around it. The cylinder is placed on a horizontal surface and the free end of the cable is held horizontally, and tugged with force F, such that the cable unwinds from the top of the cylinder. Assuming that the cylinder rolls without slipping, what is its acceleration? [Ans: $(4/3)\,F/M$.]

8.13 A uniform cylinder of radius $b = 0.25\,\mathrm{m}$ is given an angular speed of $\omega_0 = 35\,\mathrm{rad/s}$ about an axis, parallel to its length, that passes through its center. The cylinder is gently lowered onto a horizontal frictional surface, and released. The coefficient of friction of the surface is $\mu = 0.15$.

(a) How long does it take before the cylinder starts to roll without slipping? [Ans: 1.98 s.]

(b) What distance does the cylinder travel between its release point and the point at which it commences to roll without slipping? [Ans: 2.88 m.]

8.14 A solid sphere rolls without slipping down a rough slope whose inclination to the horizontal is θ. Let μ be the coefficient of friction between the sphere and the slope.

(a) What is the acceleration of the sphere down the slope? [Ans: $(5/7)\,g\,\sin\theta$.]

(b) What is the critical value of θ above which rolling without slipping is not possible? [Ans: $\tan^{-1}[(7/2)\,\mu]$.]

8.15 A hollow sphere rolls without slipping down a rough slope whose inclination to the horizontal is θ. Let μ be the coefficient of friction between the sphere and the slope.

(a) What is the acceleration of the sphere down the slope? [Ans: $(3/5)\,g\,\sin\theta$.]

(b) What is the critical value of θ above which rolling without slipping is not possible? [Ans: $\tan^{-1}[(5/2)\,\mu]$.]

8.16 A billiard ball of weight W and radius R is subject to a horizontal force, F, as shown in Figure 8.13. Let μ be the coefficient of friction between the ball and the billiard table.

(a) Assuming that $F > \mu W$, at what height, d, must the force act on the ball in order to make it slide across the table without rotating? [Ans: $(1 - \mu W/F)\,R$.]

(b) What is the acceleration of the ball? [Ans: $(F/W - \mu)\,g$.]

8.17 A billiard ball of mass M and radius R is subject to a horizontal impulse, F, as shown in Figure 8.13. At what height, d, must the impulse act on the ball in order to make it roll without slipping, without depending on friction? Treat the ball as a homogeneous sphere. [Ans: $(7/5)\,R$.]

8.18 A uniform rod of mass M is placed like a ladder with one end against a smooth vertical wall, and the other end on a smooth horizontal plane. The bar is released from rest at an inclination α to the vertical.

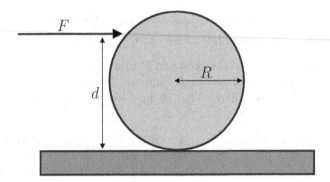

Figure 8.13 Figure for Exercise 8.16

(a) What is the initial reaction at the wall? [Ans: $(3/4)\,M\,g\,\sin\alpha\,\cos\alpha$.]

(b) What is the initial reaction at the plane? [Ans: $M\,g\,[1 - (3/4)\,\sin^2\alpha]$.]

(c) Demonstrate that the bar will cease to touch the wall when its upper end has fallen through one-third of its initial altitude.

8.19 A uniform rod is struck at rest at one end by a perpendicular impulse. Show that its kinetic energy will be greater than if the other end had been fixed, in the ratio $4 : 3$. [From Lamb 1942.]

8.20 A thin uniform plank of length l lies at rest on a horizontal sheet of ice. If the plank is given a kick at one end in a direction normal to the plank, show that the plank will begin to rotate about a point located a distance $l/6$ from its center. [From Fowles & Cassiday 2005.]

CHAPTER 9

Angular Momentum

9.1 INTRODUCTION

Two physical quantities are noticeable by their absence in Table 8.1. Namely, momentum, and its rotational concomitant, **angular momentum**. It turns out that angular momentum is a sufficiently important concept to merit a separate discussion.

9.2 ANGULAR MOMENTUM OF A POINT PARTICLE

Consider a particle of mass m, position vector $\mathbf{r}$, and instantaneous velocity $\mathbf{v}$, that rotates about an axis passing through the origin of our coordinate system. We know that the particle's linear momentum is written

$$\mathbf{p} = m\,\mathbf{v}, \tag{9.1}$$

and satisfies

$$\frac{d\mathbf{p}}{dt} = \mathbf{f}, \tag{9.2}$$

where $\mathbf{f}$ is the force acting on the particle. (See Chapter 6.) Let us search for the rotational equivalent of $\mathbf{p}$.

Consider the quantity

$$\mathbf{l} = \mathbf{r} \times \mathbf{p}. \tag{9.3}$$

This quantity, which is known as *angular momentum*, is a vector of magnitude

$$l = r\,p\,\sin\theta, \tag{9.4}$$

where θ is the angle subtended between the directions of $\mathbf{r}$ and $\mathbf{p}$. The direction of $\mathbf{l}$ is defined to be mutually perpendicular to the directions of $\mathbf{r}$ and $\mathbf{p}$, in the sense given by the right-hand grip rule. In other words, if vector $\mathbf{r}$ rotates onto vector $\mathbf{p}$ (through an angle less than 180°), and the fingers of the right-hand are aligned with this rotation, then the thumb of the right-hand indicates the direction of $\mathbf{l}$. See Figure 9.1, where $\mathbf{l}$ is directed out of the page.

Let us differentiate Equation (9.3) with respect to time. We obtain

$$\frac{d\mathbf{l}}{dt} = \dot{\mathbf{r}} \times \mathbf{p} + \mathbf{r} \times \dot{\mathbf{p}}. \tag{9.5}$$

DOI: 10.1201/9781003198642-9

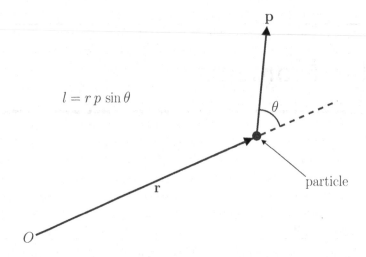

$$l = r\,p\,\sin\theta$$

Figure 9.1 Angular momentum of a point particle about the origin. The angular momentum vector is directed out of the page.

Note that the derivative of a vector product is formed in much the same manner as the derivative of an ordinary product, except that the order of the various terms is preserved. We know that $\dot{\mathbf{r}} = \mathbf{v} = \mathbf{p}/m$ and $\dot{\mathbf{p}} = \mathbf{f}$. Hence, we obtain

$$\frac{d\mathbf{l}}{dt} = \frac{\mathbf{p} \times \mathbf{p}}{m} + \mathbf{r} \times \mathbf{f}. \tag{9.6}$$

However, $\mathbf{p} \times \mathbf{p} = \mathbf{0}$, because the vector product of two parallel vectors is zero. (See Section 3.2.6.) Furthermore,

$$\mathbf{r} \times \mathbf{f} = \boldsymbol{\tau}, \tag{9.7}$$

where $\boldsymbol{\tau}$ is the torque acting on the particle about an axis passing through the origin. (See Section 8.6.) We conclude that

$$\frac{d\mathbf{l}}{dt} = \boldsymbol{\tau}. \tag{9.8}$$

Of course, given that torque is the rotational equivalent of force, this equation is analogous to Equation (9.2), which suggests that angular momentum, $\mathbf{l}$, plays the role of linear momentum, $\mathbf{p}$, in rotational dynamics.

For the special case of a particle of mass m executing a circular orbit of radius r, with instantaneous velocity v, and instantaneous angular velocity $\omega = v/r$, the magnitude of the particle's angular momentum is simply

$$l = m\,v\,r = m\,\omega\,r^2. \tag{9.9}$$

9.3 ANGULAR MOMENTUM OF AN EXTENDED OBJECT

Consider a rigid object rotating about some fixed axis with angular velocity $\boldsymbol{\omega}$. Let us model this object as a swarm of N particles. Suppose that the ith particle has mass m_i, position vector $\mathbf{r}_i$, and velocity $\mathbf{v}_i$. Incidentally, it is assumed that the object's axis of rotation passes through the origin of our coordinate system. The total angular momentum of the object, $\mathbf{L}$, is simply the vector sum of the angular momenta of its N constituent particles. Hence,

$$\mathbf{L} = \sum_{i=1,N} m_i\,\mathbf{r}_i \times \mathbf{v}_i. \tag{9.10}$$

For a rigidly rotating object, we can write

$$\mathbf{v}_i = \boldsymbol{\omega} \times \mathbf{r}_i. \tag{9.11}$$

(See Section 8.3.) Let

$$\boldsymbol{\omega} = \omega\,\mathbf{k}, \tag{9.12}$$

where $\mathbf{k}$ is a unit vector directed along the object's axis of rotation (in the sense given by the right-hand grip rule). It follows that

$$\mathbf{L} = \omega \sum_{i=1,N} m_i\,\mathbf{r}_i \times (\mathbf{k} \times \mathbf{r}_i). \tag{9.13}$$

Let us calculate the component of $\mathbf{L}$ along the object's rotation axis; that is, the component along the $\mathbf{k}$-axis. We can write

$$L_k = \mathbf{L} \cdot \mathbf{k} = \omega \sum_{i=1,N} m_i\,\mathbf{k} \cdot \mathbf{r}_i \times (\mathbf{k} \times \mathbf{r}_i). \tag{9.14}$$

However, because $\mathbf{a} \cdot \mathbf{b} \times \mathbf{c} = \mathbf{a} \times \mathbf{b} \cdot \mathbf{c}$, the previous expression can be rewritten as

$$L_k = \omega \sum_{i=1,N} m_i\,(\mathbf{k} \times \mathbf{r}_i) \cdot (\mathbf{k} \times \mathbf{r}_i) = \omega \sum_{i=1,N} m_i\,|\mathbf{k} \times \mathbf{r}_i|^2. \tag{9.15}$$

Now,

$$\sum_{i=1,N} m_i\,|\mathbf{k} \times \mathbf{r}_i|^2 = I_{kk}, \tag{9.16}$$

where I_{kk} is the moment of inertia of the object about the $\mathbf{k}$-axis. (See Section 8.5.) Hence, it follows that

$$L_k = I_{kk}\,\omega. \tag{9.17}$$

According to the previous formula, the component of a rigid object's angular momentum vector along its axis of rotation is simply the product of the body's moment of inertia about this axis and the object's angular velocity. Does this result imply that we can automatically write

$$\mathbf{L} = I\,\boldsymbol{\omega}? \tag{9.18}$$

Unfortunately, in general, the previous equation is incorrect. This conclusion follows because the object may possess non-zero angular momentum components about axes perpendicular to its axis of rotation. Thus, in general, the angular momentum vector of a rotating object is not parallel to its angular velocity vector, which constitutes a major difference from translational motion, where linear momentum is always found to be parallel to linear velocity.

For a rigid object rotating with angular velocity $\boldsymbol{\omega} = (\omega_x,\ \omega_y,\ \omega_z)$, we can write the object's angular momentum $\mathbf{L} = (L_x,\ L_y,\ L_z)$ in the form

$$L_x = I_{xx}\,\omega_x, \tag{9.19}$$

$$L_y = I_{yy}\,\omega_y, \tag{9.20}$$

$$L_z = I_{zz}\,\omega_z, \tag{9.21}$$

where I_{xx} is the moment of inertia of the object about the x-axis, etcetera. Here, it is again assumed that the origin of our coordinate system lies on the object's axis of rotation. Note that the previous equations are only valid when the x-, y-, and z-axes are aligned in a certain very special manner. In fact, the Cartesian axes must be aligned along the so-called *principal axes of rotation* of the object. (The principal axes invariably coincide with the object's main symmetry axes.) It turns out that it is always possible to find three, mutually perpendicular, principal axes of rotation that pass through a given point in a rigid object. Reconstructing **L** from its components, we obtain

$$\mathbf{L} = I_{xx}\,\omega_x\,\mathbf{e}_x + I_{yy}\,\omega_y\,\mathbf{e}_y + I_{zz}\,\omega_z\,\mathbf{e}_z, \tag{9.22}$$

where $\mathbf{e}_x$ is a unit vector pointing along the x-axis, etcetera. It is clear, from the previous equation, that the reason **L** is not generally parallel to $\boldsymbol{\omega}$ is because the moments of inertia of a rigid object about its three principal axes of rotation are not generally the same. In other words, if $I_{xx} = I_{yy} = I_{zz} = I$ then $\mathbf{L} = I\,\boldsymbol{\omega}$, and the angular momentum and angular velocity vectors are always parallel. However, if $I_{xx} \neq I_{yy} \neq I_{zz}$, which is usually the case, then **L** is not, in general, parallel to $\boldsymbol{\omega}$.

Although Equation (9.22) suggests that the angular momentum of a rigid object is not generally parallel to its angular velocity, this equation also implies that there are, at least, three special axes of rotation for which this is the case. Suppose, for instance, that the object rotates about the z-axis, so that $\boldsymbol{\omega} = \omega_z\,\mathbf{e}_z$. It follows from Equation (9.22) that

$$\mathbf{L} = I_{zz}\,\omega_z\,\mathbf{e}_z = I_{zz}\,\boldsymbol{\omega}. \tag{9.23}$$

Thus, in this case, the angular momentum vector is parallel to the angular velocity vector. The same can be said for rotation about the x- or y-axes. We conclude that when a rigid object rotates about one of its principal axes then its angular momentum is parallel to its angular velocity, but not, in general, otherwise.

How can we identify a principal axis of a rigid object? At the simplest level, a principal axis is one about which the object possesses axial symmetry. The required type of symmetry is illustrated in Figure 9.2. Assuming that the object can be modeled as a swarm of particles, for every particle of mass m, located a distance r from the origin, and subtending an angle θ with the rotation axis, there must be an identical particle located on diagrammatically the opposite side of the rotation axis. As shown in the figure, the angular momentum vectors of such a matched pair of particles can be added together to form a resultant angular momentum vector that is parallel to the axis of rotation. Thus, if the object is composed entirely of matched particle pairs then its angular momentum vector must be parallel to its angular velocity vector. The generalization of this argument to deal with continuous objects is fairly straightforward. For instance, symmetry implies that any axis of rotation that passes through the center of a uniform sphere is a principal axis of that object. Likewise, a perpendicular axis that passes through the center of a uniform disk is a principal axis. Finally, a perpendicular axis that passes through the center of a uniform rod is a principal axis.

9.4   ANGULAR MOMENTUM OF A MULTI-COMPONENT SYSTEM

Consider a system consisting of N mutually interacting point particles. Such a system might represent a true multi-component system, such as an asteroid cloud, or it might represent an extended body. Let the ith particle, whose mass is m_i, be located at vector displacement $\mathbf{r}_i$. Suppose that this particle exerts a force $\mathbf{f}_{ji}$ on the jth particle. By Newton's third law of motion, the force, $\mathbf{f}_{ij}$, exerted by the jth particle on the ith is given by

$$\mathbf{f}_{ij} = -\mathbf{f}_{ji}. \tag{9.24}$$

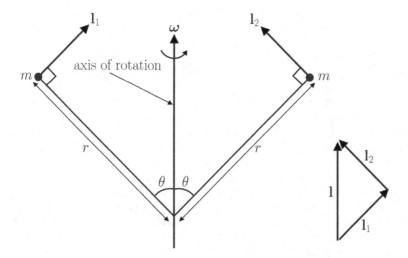

Figure 9.2 A principal axis of rotation. The velocity vectors of the left and right masses are directed out of the page and into the page, respectively.

(See Section 4.5.)

Let us assume that the internal forces acting within the system are *central forces*; that is, the force $\mathbf{f}_{ij}$, acting between particles i and j, is directed along the line of centers of these particles. In other words,

$$\mathbf{f}_{ij} \propto \mathbf{r}_i - \mathbf{r}_j. \tag{9.25}$$

See Figure 9.3. Incidentally, this is not a particularly restrictive assumption, because most forces that occur in nature are central forces. For instance, gravity is a central force, electrostatic forces are central, and the internal stresses acting within a rigid body are approximately central. Suppose, finally, that the ith particle is subject to an external force, $\mathbf{F}_i$.

The equation of motion of the ith particle can be written

$$\dot{\mathbf{p}}_i = \sum_{\substack{j=1,N}}^{j \neq i} \mathbf{f}_{ij} + \mathbf{F}_i. \tag{9.26}$$

Taking the vector product of this equation with the position vector $\mathbf{r}_i$, we obtain

$$\mathbf{r}_i \times \dot{\mathbf{p}}_i = \sum_{\substack{j=1,N}}^{j \neq i} \mathbf{r}_i \times \mathbf{f}_{ij} + \mathbf{r}_i \times \mathbf{F}_i. \tag{9.27}$$

Now,

$$\mathbf{r}_i \times \dot{\mathbf{p}}_i = \frac{d(\mathbf{r}_i \times \mathbf{p}_i)}{dt}, \tag{9.28}$$

because $\dot{\mathbf{r}}_i \propto \mathbf{p}_i$, and $\mathbf{p}_i \times \mathbf{p}_i = \mathbf{0}$. We also know that the total angular momentum, $\mathbf{L}$, of the system (about the origin) can be written in the form

$$\mathbf{L} = \sum_{i=1,N} \mathbf{r}_i \times \mathbf{p}_i. \tag{9.29}$$

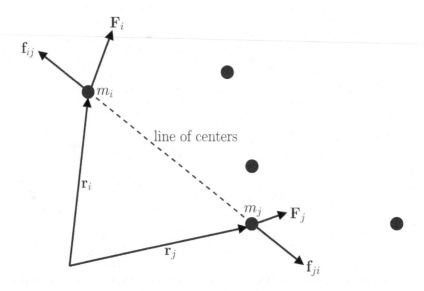

Figure 9.3 A multi-component system with central internal forces

Hence, summing Equation (9.27) over all particles, we obtain

$$\frac{d\mathbf{L}}{dt} = \sum_{i,j=1,N}^{i \neq j} \mathbf{r}_i \times \mathbf{f}_{ij} + \sum_{i=1,N} \mathbf{r}_i \times \mathbf{F}_i. \tag{9.30}$$

Consider the first expression on the right-hand side of Equation (9.30). A general term, $\mathbf{r}_i \times \mathbf{f}_{ij}$, in this sum can always be paired with a matching term, $\mathbf{r}_j \times \mathbf{f}_{ji}$, in which the indices have been swapped. Making use of Equation (9.24), the sum of a general matched pair can be written

$$\mathbf{r}_i \times \mathbf{f}_{ij} + \mathbf{r}_j \times \mathbf{f}_{ji} = (\mathbf{r}_i - \mathbf{r}_j) \times \mathbf{f}_{ij}. \tag{9.31}$$

However, if the internal forces are central in nature then $\mathbf{f}_{ij}$ is parallel to $\mathbf{r}_i - \mathbf{r}_j$. Hence, the vector product of these two vectors is zero. (See Section 3.2.6.) We conclude that

$$\mathbf{r}_i \times \mathbf{f}_{ij} + \mathbf{r}_j \times \mathbf{f}_{ji} = \mathbf{0}, \tag{9.32}$$

for any values of i and j (provided $i \neq j$). Thus, the first term on the right-hand side of Equation (9.30) sums to zero. We are left with

$$\frac{d\mathbf{L}}{dt} = \boldsymbol{\tau}, \tag{9.33}$$

where

$$\boldsymbol{\tau} = \sum_{i=1,N} \mathbf{r}_i \times \mathbf{F}_i \tag{9.34}$$

is the net external torque acting on the system (about an axis passing through the origin). Of course, Equation (9.33) is simply the rotational equation of motion for the system taken as a whole.

Suppose that the system is isolated, such that it is subject to zero net external torque. It follows from Equation (9.33) that, in this case, the total angular momentum of the system

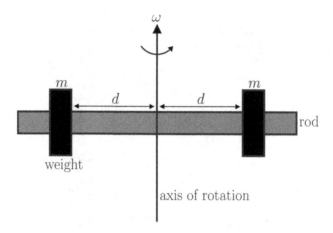

Figure 9.4 Two movable weights on a rotating rod

is a conserved quantity. To be more exact, the components of the total angular momentum taken about any three independent axes are individually conserved quantities. Conservation of angular momentum is an extremely useful concept that greatly simplifies the analysis of a wide range of rotating systems. Let us consider some examples.

9.5 CONSERVATION OF ANGULAR MOMENTUM

9.5.1 Two Movable Weights on a Rotating Rod

Suppose that two identical weights of mass m are attached to a light rigid rod that rotates without friction about a perpendicular axis passing through its mid-point. Imagine that the two weights are equipped with small motors that allow them to travel along the rod; the motors are synchronized in such a manner that the distance of the two weights from the axis of rotation is always the same. Let us call this common distance d and let ω be the angular velocity of the rod. See Figure 9.4. How does the angular velocity ω change as the distance d is varied?

Note that there are no external torques acting on the system. It follows that the system's angular momentum must remain constant as the weights move along the rod. Neglecting the contribution of the rod, the moment of inertia of the system is written

$$I = 2\,m\,d^2. \tag{9.35}$$

Because the system is rotating about a principal axis, its angular momentum takes the form

$$L = I\,\omega = 2\,m\,d^2\,\omega. \tag{9.36}$$

If L is a constant of the motion then we obtain

$$\omega\,d^2 = \text{constant}. \tag{9.37}$$

In other words, the system spins faster as the weights move inward toward the axis of rotation, and vice versa.

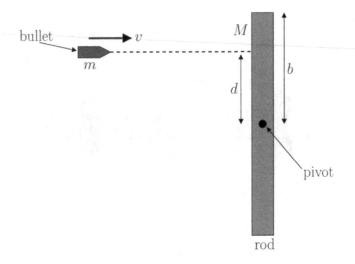

Figure 9.5 A bullet strikes a pivoted rod

9.5.2 Figure Skater

The effect described in the previous subsection is familiar from figure skating. When a female (say) skater spins about a vertical axis, her angular momentum is approximately a conserved quantity, because the ice exerts very little torque on her. Thus, if the skater starts spinning with outstretched arms, and then draws her arms inward, then her moment of inertia deceases, and her rate of rotation spontaneously increases in order to conserve angular momentum. The skater can slow her rate of rotation by simply pushing her arms outward again.

9.5.3 Bullet Striking a Pivoted Rod

Suppose that a bullet of mass m and velocity v strikes, and becomes embedded in, a stationary rod of mass M and length $2\,b$ that pivots about a frictionless perpendicular axle passing through its mid-point. Let the bullet strike the rod normally a distance d from its axis of rotation. See Figure 9.5. What is the instantaneous angular velocity ω of the rod (and bullet) immediately after the collision?

Taking the bullet and the rod as a whole, this is again a system upon which no external torque acts. Thus, we expect the system's net angular momentum to be the same before and after the collision. Before the collision, only the bullet possesses angular momentum, because the rod is at rest. As is easily demonstrated, the bullet's angular momentum about the pivot point is

$$l = m\,v\,d; \tag{9.38}$$

that is, the product of its mass, its velocity, and its distance of closest approach to the point about which the angular momentum is measured. The previous expression is a general result (for a point particle). After the collision, the bullet lodges a distance d from the pivot, and is forced to co-rotate with the rod. Hence, the angular momentum of the bullet after the collision is given by

$$l' = m\,d^2\,\omega, \tag{9.39}$$

where ω is the angular velocity of the rod. The angular momentum of the rod after the collision is

$$L = I\omega, \tag{9.40}$$

where $I = (1/12)\,M\,(2\,b)^2 = (1/3)\,M\,b^2$ is the rod's moment of inertia (about a perpendicular axis passing through its mid-point). (See Section 8.5.4.) Conservation of angular momentum yields

$$l = l' + L, \tag{9.41}$$

or

$$\omega = \frac{m\,v\,d}{I + m\,d^2}. \tag{9.42}$$

The kinetic energy of the system before the collision is

$$K_i = \frac{1}{2}\,m\,v^2. \tag{9.43}$$

The kinetic energy of the system immediately after the collision is

$$K_f = \frac{1}{2}\,m\,d^2\,\omega^2 + \frac{1}{2}\,I\,\omega^2. \tag{9.44}$$

Hence, the fractional loss in kinetic energy in the collision is

$$\frac{K_i - K_f}{K_i} = \frac{I}{I + m\,d^2}. \tag{9.45}$$

It can be seen that there is always an energy loss associated with the type of totally inelastic collision considered in this subsection. In fact, if $m\,d^2 \ll I$, as is likely to be the case, then the fractional energy loss is very significant.

9.6   SPINNING TOP

Consider the spinning top pictured in Figure 9.6. The top pivots on the ground, while spinning about its axis at angular velocity ω. Let the spin axis subtend an angle θ with the vertical. Let M be the mass of the top, d the distance of the top's center of mass from the pivot point, and $I_\parallel$ the top's moment of inertia about its spin axis (which is assumed to be a principal axis). There are two forces acting on the top. First, the top's weight, which is of magnitude $M\,g$, and acts vertically downward at the center of mass. Second, the reaction at the pivot, which is also of magnitude $M\,g$, and acts vertically upward. The two forces balance one another in the vertical direction. However, the top's weight exerts a horizontal torque about the pivot point of magnitude

$$\tau = M\,g\,d\,\sin\theta. \tag{9.46}$$

The direction of the torque is into the page in the side-view part of Figure 9.6. This torque acts to twist the top downward. If the top were not spinning then the torque would cause the top to fall to the ground. However, it is a matter of experience that if such a torque acts on a rapidly spinning top then the top does not fall over; instead it *precesses* about a vertical axis passing through the pivot point. Let us investigate this effect.

Let us setup the Cartesian coordinate system shown in Figure 9.6. The origin of the coordinate system coincides with the pivot point. The z-axis points vertically upward, and the x- and y-axes lie in the horizontal plane. Suppose that the projection of the top's spin angular velocity vector onto the x-y plane subtends a counter-clockwise (looking down)

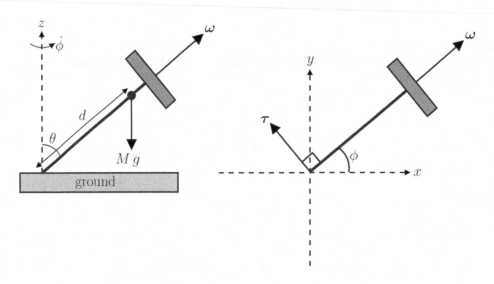

Figure 9.6   A spinning top

angle ϕ with the x-axis. The top's spin angular velocity vector, thus, has the Cartesian components

$$\boldsymbol{\omega} = \omega \, (\sin\theta \, \cos\phi, \, \sin\theta \, \sin\phi, \, \cos\theta). \tag{9.47}$$

Hence, the top's angular momentum vector has the components

$$\mathbf{L} = I_{\parallel} \, \boldsymbol{\omega} = I_{\parallel} \, \omega \, (\sin\theta \, \cos\phi, \, \sin\theta \, \sin\phi, \, \cos\theta). \tag{9.48}$$

Now, the torque vector lies in the x-y plane, and subtends a right-angle with the projection of $\boldsymbol{\omega}$ onto this plane, as illustrated in Figure 9.6. Thus, we can write

$$\boldsymbol{\tau} = \tau \, (-\sin\phi, \, \cos\phi, \, 0) = M \, g \, d \, (-\sin\theta \, \sin\phi, \, \sin\theta \, \cos\phi, \, 0). \tag{9.49}$$

The top's equation of angular motion is

$$\frac{d\mathbf{L}}{dt} = \boldsymbol{\tau}. \tag{9.50}$$

Let us assume that the top's motion consists of pure precession. In this case, the only time-varying quantity is the azimuthal angle, ϕ. Hence, according to Equation (9.48),

$$\frac{d\mathbf{L}}{dt} = I_{\parallel} \, \omega \, \dot{\phi} \, (-\sin\theta \, \sin\phi, \, \sin\theta \, \cos\phi, \, 0). \tag{9.51}$$

The previous three equations can be combined to give

$$I_{\parallel} \, \omega \, \dot{\phi} \, (-\sin\theta \, \sin\phi, \, \sin\theta \, \cos\phi, \, 0) = M \, g \, d \, (-\sin\theta \, \sin\phi, \, \sin\theta \, \cos\phi, \, 0), \tag{9.52}$$

which can be satisfied provided

$$\dot{\phi} = \Omega_\phi, \tag{9.53}$$

where

$$\Omega_\phi = \frac{M g d}{I_\parallel \, \omega}. \qquad (9.54)$$

It follows that the gravitational torque acting on the top causes the azimuthal angle ϕ to increase in time at the constant rate Ω_ϕ. As is clear from the top-view part of Figure 9.6, this implies that the top precesses about a vertical axis passing through the pivot point with angular velocity Ω_ϕ. The direction of the precession is the same as the (vertical projection of) the top's direction of spin about its axis.

The previous analysis is only valid provided

$$\Omega_\phi \ll \omega; \qquad (9.55)$$

that is, provided the precession rate is small compared to the top's spin rate. The reason for this is that we completely neglected the contribution of the precession to the top's angular momentum. As can be seen from Equation (9.54), the previous constraint can always be satisfied provided the top's spin rate is made sufficiently large.

The top's gravitational potential energy is written

$$U = M g d \, \cos \theta. \qquad (9.56)$$

Note that

$$\tau = -\frac{dU}{d\theta} = M g d \, \sin \theta. \qquad (9.57)$$

See Equation (9.46). The previous equation is the rotational equivalent of Equation (5.43).

It may seem that we have spent an inordinate amount of time discussing the physics of a children's toy. However, it turns out that the Earth's itself acts like a spinning top, due to its diurnal rotation. Moreover, because the Earth is slightly flattened at its poles (see Section 12.3), it is subject to a gravitational torque from the Sun and the Moon. This torque causes the Earth's axis of rotation to slowly precess about an axis that is perpendicular to the plane of its orbit around the Sun. As we shall see in Section 15.8, the physics of this so-called *luni-solar precession* is almost identical to the physics of the precession of a spinning top.

9.7   EXERCISES

9.1 A missile of mass $m = 2.3 \times 10^4$ kg flies level to the ground at an altitude of $d = 10\,000$ m with constant speed $v = 210$ m/s. What is the magnitude of the missile's angular momentum relative to a point on the ground directly below its flight path? [Ans: 4.83×10^{10} kg m^2/s.]

9.2 A uniform sphere of mass $M = 5$ kg and radius $a = 0.2$ m spins about an axis passing through its center with period $T = 0.7$ s. What is the angular momentum of the sphere? [Ans: 0.718 kg m^2/s.]

9.3 A skater spins at an initial angular velocity of 11 rad/s with arms outstretched. The skater then lowers her/his arms, thereby decreasing her/his moment of inertia by a factor 8. What is the skater's final angular velocity? Assume that any friction between the skater's skates and the ice is negligible. [Ans: 88 rad/s.]

9.4 A flywheel rotates without friction in a horizontal plane at angular velocity ω. A second flywheel, which is at rest, and has a moment of inertia three times that of the rotating flywheel, is dropped onto the rotating flywheel. As a consequence of friction, the two flywheels quickly attain the same angular velocity.

(a) What is the final angular velocity of the system? [Ans: $\omega/4$.]

(b) What fraction of the initial kinetic energy is lost in the coupling of the flywheels? [Ans: 3/4.]

9.5 A bullet of mass m and velocity v strikes the edge of a solid disk of mass M and radius b tangentially, and becomes embedded in the disk. The initial trajectory of the bullet lies in the plane of the disk. The disk is free to rotate about a perpendicular axis that passes through its center, and is initially at rest. What is the angular velocity of the disk immediately after the bullet becomes embedded? [Ans: $(v/b)/(1+M/2\,m)$.]

CHAPTER 10

Statics

10.1 INTRODUCTION

One of the most useful application of the laws of mechanics is the study of situations in which nothing moves; this discipline is known as **statics**. The principles of statics are employed by engineers whenever they design stationary structures, such as buildings, bridges, and tunnels, in order to ensure that these structures do not collapse.

10.2 PRINCIPLES OF STATICS

Consider a general extended body that is subject to a number of external forces. Let us model this body as a swarm of N point particles. In the limit that $N \to \infty$, such a model becomes a fully accurate representation of the body's dynamics.

In Section 6.3, we determined that the overall translational equation of motion of a general N-component system can be written in the form

$$\frac{d\mathbf{P}}{dt} = \mathbf{F}. \tag{10.1}$$

Here, $\mathbf{P}$ is the total linear momentum of the system, and

$$\mathbf{F} = \sum_{i=1,N} \mathbf{F}_i \tag{10.2}$$

is the resultant of all of the external forces acting on the system. Note that $\mathbf{F}_i$ is the external force acting on the ith component of the system.

Equation (10.1) effectively determines the translational motion of the system's center of mass. However, in order to fully determine the motion of the system, we must also follow its rotational motion about its center of mass (or any other convenient reference point). In Section 9.4, we determined that the overall rotational equation of motion of a general N-component system (with central internal forces) can be written in the form

$$\frac{d\mathbf{L}}{dt} = \boldsymbol{\tau}. \tag{10.3}$$

Here, $\mathbf{L}$ is the total angular momentum of the system (about the origin of our coordinate system), and

$$\boldsymbol{\tau} = \sum_{i=1,N} \mathbf{r}_i \times \mathbf{F}_i \tag{10.4}$$

DOI: 10.1201/9781003198642-10

is the resultant of all of the external torques acting on the system (about the origin of our coordinate system). In the previous equation, $\mathbf{r}_i$ is the vector displacement of the ith component of the system.

What conditions must be satisfied by the various external forces and torques acting on the system if it is to remain in a stationary state? Clearly, if the system is initially stationary then its net linear momentum, $\mathbf{P}$, and its net angular momentum, $\mathbf{L}$, are both zero. In order for the system to remain stationary, $\mathbf{P}$ and $\mathbf{L}$ must both remain constant in time. In other words, $d\mathbf{P}/dt = d\mathbf{L}/dt = \mathbf{0}$. It follows from Equations (10.1) and (10.3) that

$$\mathbf{F} = \mathbf{0}, \tag{10.5}$$

$$\boldsymbol{\tau} = \mathbf{0}. \tag{10.6}$$

In other words, the net external force acting on system must be zero, and the net external torque acting on the system must be zero. To be more exact:

> The components of the net external force acting along any three independent directions must all be zero.

and

> The components of the net external torques acting about any three independent axes (passing through the origin of the coordinate system) must all be zero.

In a nutshell, these are the principles of statics.

It is clear that the previous principles are necessary conditions for a general physical system that is initially stationary not to evolve in time. But, are they also sufficient conditions? In other words, is it necessarily true that a general system that satisfies these conditions does not exhibit any time variation? The answer to this question is as follows. If the system under investigation is a rigid body, such that the motion of any component of the body necessarily implies the motion of the whole body, then the previous principles are necessary and sufficient conditions for the existence of an equilibrium state. On the other hand, if the system is not a rigid body, so that some components of the body can move independently of others, then the previous conditions only guarantee that the system remains static in an average sense.

Before we attempt to apply the principles of statics, there are a few important points that need clarification. First, does it matter about which point we calculate the net torque acting on the system? To be more exact, if we determine that the net torque acting about a given point is zero then does this necessarily imply that the net torque acting about any other point is also zero? Now,

$$\boldsymbol{\tau} = \sum_{i=1,N} \mathbf{r}_i \times \mathbf{F}_i \tag{10.7}$$

is the net torque acting on the system about the origin of our coordinate system. The net torque about some general point, $\mathbf{r}_0$, is simply

$$\boldsymbol{\tau}' = \sum_{i=1,N} (\mathbf{r}_i - \mathbf{r}_0) \times \mathbf{F}_i. \tag{10.8}$$

However, we can rewrite the previous expression as

$$\boldsymbol{\tau}' = \sum_{i=1,N} \mathbf{r}_i \times \mathbf{F}_i - \mathbf{r}_0 \times \left(\sum_{i=1,N} \mathbf{F}_i \right) = \boldsymbol{\tau} + \mathbf{r}_0 \times \mathbf{F}. \tag{10.9}$$

If the system is in equilibrium then $\mathbf{F} = \boldsymbol{\tau} = \mathbf{0}$. Hence, it follows from the previous equation that

$$\boldsymbol{\tau}' = \mathbf{0}. \tag{10.10}$$

In other words, if the system is in equilibrium then the determination that the net torque acting about a given point is zero necessarily implies that the net torque acting about any other point is also zero. Hence, we can choose the point about which we calculate the net torque at will; this choice is usually made so as to simplify the calculation.

Another question that needs clarification is as follows. At which point should we assume that the weight of the system acts in order to calculate the contribution of the weight to the net torque acting about a given point? Actually, in Section 8.11, we effectively answered this question by assuming that the weight acts at the center of mass of the system. Let us now justify this assumption. The external force acting on the ith component of the system due to its weight is

$$\mathbf{F}_i = m_i\,\mathbf{g}, \tag{10.11}$$

where $\mathbf{g}$ is the acceleration due to gravity (which is assumed to be uniform throughout the system). Hence, the net gravitational torque acting on the system about the origin of our coordinate scheme is

$$\boldsymbol{\tau} = \sum_{i=1,N} \mathbf{r}_i \times m_i\,\mathbf{g} = \left(\sum_{i=1,N} m_i\,\mathbf{r}_i \right) \times \mathbf{g} = \mathbf{r}_{cm} \times M\,\mathbf{g}, \tag{10.12}$$

where $M = \sum_{i=1,N} m_i$ is the total mass of the system and $\mathbf{r}_{cm} = \sum_{i=1,N} m_i\,\mathbf{r}_i/M$ is the position vector of its center of mass. It follows, from the previous equation, that the net gravitational torque acting on the system about a given point can be calculated by assuming that the total mass of the system is concentrated at its center of mass.

10.3 EQUILIBRIUM OF A LAMINAR OBJECT

Consider a general laminar object that is free to pivot about a fixed horizontal axis perpendicular to its plane. Assuming that the object is placed in a uniform gravitational field (such as that on the surface of the Earth), what is the object's equilibrium configuration in this field? Let O represent the pivot point, and let C be the center of mass of the object. See Figure 10.1. Suppose that r represents the distance between points O and C, whereas θ is the angle subtended between the line OC and the downward vertical. There are two external forces acting on the object. First, there is the downward force, $M\,g$, due to gravity, which acts at the center of mass. Second, there is the reaction, R, due to the pivot, which acts at the pivot point. Here, M is the mass of the object, and g is the acceleration due to gravity.

Two conditions must be satisfied in order for a given configuration of the object shown in Figure 10.1 to represent an equilibrium configuration. First, there must be zero net external force acting on the object. This implies that the reaction, R, is equal and opposite to the gravitational force, $M\,g$. In other words, the reaction is of magnitude $M\,g$, and is directed vertically upward. The second condition is that there must be zero net torque acting about the pivot point. Now, the reaction, R, does not generate a torque, because it acts at the pivot point. Moreover, the torque associated with the gravitational force, $M\,g$, is simply the magnitude of this force times the length of the lever arm, d. (See Figure 10.1.) Hence, the net torque acting on the system about the pivot point is

$$\tau = M\,g\,d = M\,g\,r\,\sin\theta. \tag{10.13}$$

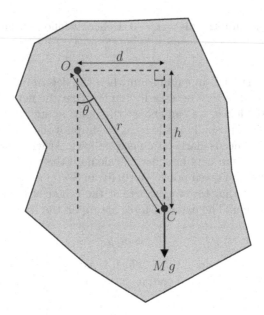

Figure 10.1 A laminar object pivoting about a fixed point in a gravitational field

Setting this torque to zero, we obtain $\sin\theta = 0$, which implies that $\theta = 0$. In other words, the equilibrium configuration of a general laminar object (that is free to rotate about a fixed horizontal axis perpendicular to its plane in a uniform gravitational field) is that in which the center of mass of the object is aligned vertically below the pivot point.

Incidentally, we can use the previous result to experimentally determine the center of mass of a given laminar object. We would need to successively suspend the object from two different pivot points. In each equilibrium configuration, we would then need to mark a line running vertically downward from the pivot point, using a plumb-line. The crossing point of these two lines would indicate the position of the center of mass.

Our discussion of the equilibrium configuration of the laminar object shown in Figure 10.1 is not quite complete. We have determined that the condition which must be satisfied by an equilibrium state is $\sin\theta = 0$. However, there are, in fact, two physical roots of this equation. The first, $\theta = 0$, corresponds to the case where the center of mass of the object is aligned vertically below the pivot point. The second, $\theta = \pi$, corresponds to the case where the center of mass is aligned vertically above the pivot point. Of course, the former root is far more important than the latter, because the former root corresponds to a stable equilibrium, whereas the latter corresponds to an unstable equilibrium. Recall, from Section 5.7, that when a system is slightly disturbed from a stable equilibrium then the forces and torques that act upon it tend to return it to this equilibrium, and vice versa for an unstable equilibrium. The easiest way to distinguish between stable and unstable equilibria, in the present case, is to evaluate the gravitational potential energy of the system. The potential energy of the object shown in Figure 10.1, calculated using the height of the pivot as the reference height, is simply

$$U = -Mgh = -Mgr\cos\theta. \tag{10.14}$$

(Note that the gravitational potential energy of an extended object can be calculated by imagining that all of the mass of the object is concentrated at its center of mass.) It can be seen that $\theta = 0$ corresponds to a minimum of this potential, whereas $\theta = \pi$ corresponds to a maximum. This is in accordance with the analysis of Section 5.7, where it was demonstrated that whenever an object moves in a conservative force-field (such as a gravitational field), the stable equilibrium points correspond to minima of the potential energy associated with this field, whereas the unstable equilibrium points correspond to maxima.

10.4 RODS AND CABLES

10.4.1 Horizontal Rod Suspended from Two Cables

Consider a uniform rod of mass M and length l that is suspended horizontally via two vertical cables. Let the points of attachment of the two cables be located distances x_1 and x_2 from one of the ends of the rod, labeled A. It is assumed that $x_2 > x_1$. See Figure 10.2. What are the tensions, T_1 and T_2, in the cables?

Let us first locate the center of mass of the rod, which is situated at the rod's mid-point, a distance $l/2$ from reference point A. (See Figure 10.2.) There are three forces acting on the rod; the gravitational force, $M g$, and the two tension forces, T_1 and T_2. Each of these forces is directed vertically. Thus, the condition that zero net force acts on the system reduces to the condition that the net vertical force is zero, which yields

$$T_1 + T_2 - M g = 0. \tag{10.15}$$

Consider the torques exerted by the three previously mentioned forces about point A. Each of these torques attempts to twist the rod about an axis perpendicular to the plane of the diagram. Hence, the condition that zero net torque acts on the system reduces to the condition that the net torque at point A, about an axis perpendicular to the plane of the diagram, is zero. The contribution of each force to this torque is simply the product of the magnitude of the force and the length of the associated lever arm. In each case, the length of the lever arm is equivalent to the distance of the point of action of the force from A, measured along the length of the rod. Hence, setting the net torque to zero, we obtain

$$x_1 T_1 + x_2 T_2 - \frac{l}{2} M g = 0. \tag{10.16}$$

Note that the torque associated with the gravitational force, $M g$, has a minus sign in front, because this torque obviously attempts to twist the rod in the opposite direction to the torques associated with the tensions in the cables.

The previous two equations can be solved to give

$$T_1 = \left(\frac{x_2 - l/2}{x_2 - x_1} \right) M g, \tag{10.17}$$

$$T_2 = \left(\frac{l/2 - x_1}{x_2 - x_1} \right) M g. \tag{10.18}$$

Recall that tensions in flexible cables can never be negative, because this would imply that the cables in question were being compressed. Of course, when cables are compressed they simply collapse. It is clear, from the previous expressions, that in order for the tensions T_1 and T_2 to remain positive (given that $x_2 > x_1$), the following conditions must be satisfied:

$$x_1 < \frac{l}{2}, \tag{10.19}$$

$$x_2 > \frac{l}{2}. \tag{10.20}$$

In other words, the attachment points of the two cables must straddle the center of mass of the rod.

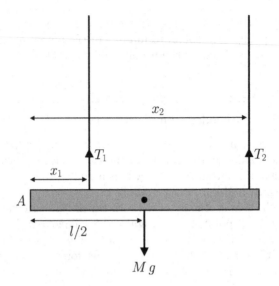

Figure 10.2 A horizontal rod suspended by two vertical cables

10.4.2 Pivoting Horizontal Rod Supported by a Cable

Consider a uniform rod of mass M and length l that is free to rotate in the vertical plane about a fixed pivot attached to one of its ends. The other end of the rod is attached to a fixed cable. We can imagine that both the pivot and the cable are anchored in the same vertical wall. See Figure 10.3. Suppose that the rod is horizontal and that the cable subtends an angle θ with the horizontal. Assuming that the rod is in equilibrium, what is the magnitude of the tension, T, in the cable, and what is the direction and magnitude of the reaction, R, at the pivot?

As usual, the center of mass of the rod lies at its mid-point. There are three forces acting on the rod; the reaction, R; the weight, $M\,g$; and the tension, T. The reaction acts at the pivot. Let ϕ be the angle subtended by the reaction with the horizontal, as shown in Figure 10.3. The weight acts at the center of mass of the rod, and is directed vertically downward. Finally, the tension acts at the end of the rod, and is directed along the cable.

Resolving horizontally, and setting the net horizontal force acting on the rod to zero, we obtain

$$R \cos\phi - T \cos\theta = 0. \tag{10.21}$$

Likewise, resolving vertically, and setting the net vertical force acting on the rod to zero, we obtain

$$R \sin\phi + T \sin\theta - M\,g = 0. \tag{10.22}$$

The previous constraints are sufficient to ensure that zero net force acts on the rod.

Let us evaluate the net torque acting at the pivot point (about an axis perpendicular to the plane of the diagram). The reaction, R, does not contribute to this torque, because it acts at the pivot point. The length of the lever arm associated with the weight, $M\,g$, is $l/2$. Simple trigonometry reveals that the length of the lever arm associated with the tension, T, is $l \sin\theta$. Hence, setting the net torque about the pivot point to zero, we obtain

$$M\,g\,\frac{l}{2} - T\,l \sin\theta = 0. \tag{10.23}$$

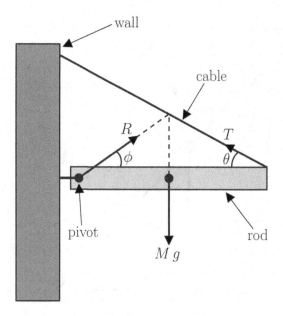

Figure 10.3 A rod suspended by a fixed pivot and a cable

Note that there is a minus sign in front of the second torque, because this torque clearly attempts to twist the rod in the opposite sense to the first.

Equations (10.21) and (10.22) can be solved to give

$$T = \frac{\cos\phi}{\sin(\theta + \phi)} M g, \tag{10.24}$$

$$R = \frac{\cos\theta}{\sin(\theta + \phi)} M g. \tag{10.25}$$

Substituting Equation (10.24) into Equation (10.23), we obtain

$$\sin(\theta + \phi) = 2 \sin\theta \cos\phi. \tag{10.26}$$

The physical solution of this equation is $\phi = \theta$ (recall that $\sin 2\theta = 2 \sin\theta \cos\theta$), which determines the direction of the reaction at the pivot. Finally, Equations (10.24) and (10.25) yield

$$T = R = \frac{M g}{2 \sin\theta}, \tag{10.27}$$

which determines both the magnitude of the tension in the cable and that of the reaction at the pivot.

One important thing to note about the previous solution is that if $\phi = \theta$ then the lines of action of the three forces—R, $M g$, and T—intersect at the same point, as shown in Figure 10.3. This is an illustration of a general rule. Namely, if a rigid body is in equilibrium under the action of three forces then these forces are either mutually parallel, as shown in Figure 10.2, or their lines of action pass through the same point, as shown in Figure 10.3. See Exercise 10.6.

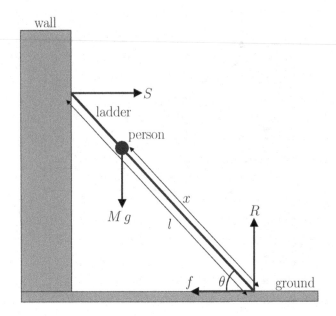

Figure 10.4 A ladder leaning against a vertical wall

10.5 LADDERS AND WALLS

Suppose that a ladder of length l and negligible mass is leaning against a vertical wall, making an angle θ with the horizontal. A person of mass M climbs a distance x along the ladder, measured from the bottom. See Figure 10.4. Suppose that the wall is completely frictionless, but that the ground possesses a coefficient of static friction μ. How far up the ladder can the person climb before it slips along the ground? Is it possible for the person to climb to the top of the ladder without any slippage occurring?

There are four forces acting on the ladder; the weight, $M g$, of the person; the reaction, S, at the wall; the reaction, R, at the ground; and the frictional force, f, due to the ground. The weight acts at the position of the person, and is directed vertically downward. The reaction, S, acts at the top of the ladder, and is directed horizontally (i.e., normal to the surface of the wall). The reaction, R, acts at the bottom of the ladder, and is directed vertically upward (i.e., normal to the ground). Finally, the frictional force, f, also acts at the bottom of the ladder, and is directed horizontally.

Resolving horizontally, and setting the net horizontal force acting on the ladder to zero, we obtain

$$S - f = 0. \tag{10.28}$$

Resolving vertically, and setting the net vertically force acting on the ladder to zero, we get

$$R - M g = 0. \tag{10.29}$$

Evaluating the torque acting about the point where the ladder touches the ground, we note that only the forces $M g$ and S contribute. The lever arm associated with the force $M g$ is $x \cos \theta$. The lever arm associated with the force S is $l \sin \theta$. Furthermore, the torques associated with these two forces act in opposite directions. Hence, setting the net torque about the bottom of the ladder to zero, we obtain

$$M g x \cos \theta - S l \sin \theta = 0. \tag{10.30}$$

The previous three equations can be solved to give

$$R = M g, \tag{10.31}$$

and

$$f = S = \frac{x}{l \tan \theta} M g. \tag{10.32}$$

The condition for the ladder not to slip with respect to the ground is (see Section 4.10)

$$f < \mu R, \tag{10.33}$$

which reduces to

$$x < l \mu \tan \theta. \tag{10.34}$$

Thus, the farthest distance that the person can climb along the ladder before it slips is

$$x_{\max} = l \mu \tan \theta. \tag{10.35}$$

Note that if $\tan \theta > 1/\mu$ then the person can climb all the way along the ladder without any slippage occurring. This result suggests that ladders leaning against walls are less likely to slip when they are almost vertical (i.e., when $\theta \to \pi/2$).

10.6 JOINTED RODS

Suppose that three identical uniform rods of mass M and length l are joined together to form an equilateral triangle, and are then suspended from a cable, as shown in Figure 10.5. What is the tension in the cable, and what are the reactions at the joints?

Let X_1, X_2, and X_3 be the horizontal reactions at the three joints, and let Y_1, Y_2, and Y_3 be the corresponding vertical reactions, as shown in Figure 10.5. In drawing this diagram, we have made use of the fact that the rods exert equal and opposite reactions on one another, in accordance with Newton's third law of motion. Let T be the tension in the cable.

Setting the horizontal and vertical forces acting on rod AB to zero, we obtain

$$X_1 - X_3 = 0, \tag{10.36}$$

$$T + Y_1 + Y_3 - M g = 0, \tag{10.37}$$

respectively. Setting the horizontal and vertical forces acting on rod AC to zero, we obtain

$$X_2 - X_1 = 0, \tag{10.38}$$

$$Y_2 - Y_1 - M g = 0, \tag{10.39}$$

respectively. Finally, setting the horizontal and vertical forces acting on rod BC to zero, we obtain

$$X_3 - X_2 = 0, \tag{10.40}$$

$$-Y_2 - Y_3 - M g = 0, \tag{10.41}$$

respectively. Incidentally, it is clear, from symmetry, that $X_1 = X_3$ and $Y_1 = Y_3$. Thus, the previous equations can be solved to give

$$T = 3 M g, \tag{10.42}$$

$$Y_2 = 0, \tag{10.43}$$

$$X_1 = X_2 = X_3 = X, \tag{10.44}$$

$$Y_1 = Y_3 = -M g. \tag{10.45}$$

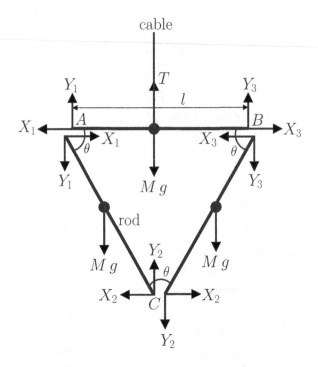

Figure 10.5 Three identical jointed rods

There only remains one unknown, X.

It is clear, from symmetry, that there is zero net torque acting on rod AB. Let us evaluate the torque acting on rod AC about point A. (By symmetry, this is the same as the torque acting on rod BC about point B). The two forces that contribute to this torque are the weight, $M g$, and the reaction $X_2 = X$. (Recall that the reaction Y_2 is zero). The lever arms associated with these two torques (which nominally act in the same direction) are $(l/2) \cos \theta$ and $l \sin \theta$, respectively. Thus, setting the net torque to zero, we obtain

$$M g \,(l/2) \cos \theta + X l \sin \theta = 0, \tag{10.46}$$

which yields

$$X = -\frac{M g}{2 \tan \theta} = -\frac{M g}{2 \sqrt{3}}, \tag{10.47}$$

because $\theta = \pi/3$ and $\tan(\pi/3) = \sqrt{3}$. We have now fully determined the tension in the cable, and all the reactions at the joints.

10.7   TIPPING OR SLIDING?

The upper part of Figure 10.6 shows a uniform rectangular block of base-length b and height h resting on a rough slope whose inclination to the horizontal is θ. Let M be the mass of the block, and let μ be the coefficient of friction between the block and the slope. Let us try to answer the following question. If the angle of the slope is gradually increased will the block eventually slide down the slope, or will it first tip over about its forward edge?

Consider the lower part of Figure 10.6, which shows the block in more detail. In this subfigure, we have reoriented our axes such that the horizontal axis runs up the slope, and

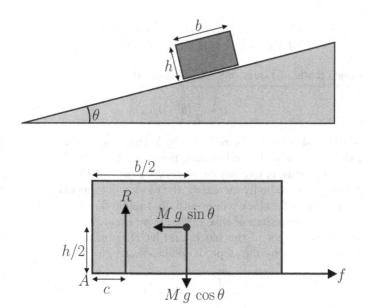

Figure 10.6 Equilibrium of a block on a rough inclined plane.

the vertical axis runs perpendicular to the slope. The weight of the block acts at the center of mass, which, by symmetry, lies at the centroid of the block. The weight can be resolved into a component $Mg \cos\theta$ acting normally into the slope, and a component $Mg \sin\theta$ acting down the slope. The line of action of the friction force, f, runs up the surface of the slope, as shown. Finally, the resultant, R, of the reaction forces acting between the block and the slope acts normally out of the slope, and has a line of action whose closest distance to the forward edge of the block is c, as shown. Because the reaction forces are all positive (i.e., the slope can only push outward on the block; it cannot pull inward) it follows that the resultant reaction cannot act outside the base of the block. In other words, the length c cannot be negative (otherwise, R would act in front of the block's base), neither can it exceed b (otherwise, R would act behind the block's base).

Let us assume that the block is in equilibrium, but is just about to start sliding down the slope. Force balance normal to the slope yields

$$R = Mg \cos\theta. \tag{10.48}$$

Force balance parallel to the slope gives

$$f = Mg \sin\theta. \tag{10.49}$$

However, if the block is just about to start sliding then

$$f = \mu R, \tag{10.50}$$

which implies that

$$\sin\theta = \mu \cos\theta. \tag{10.51}$$

Setting the net torque acting about the forward edge, A, of the block to zero, we get

$$Mg \sin\theta \frac{h}{2} + Rc = Mg \cos\theta \frac{b}{2}, \tag{10.52}$$

or

$$M g \mu \cos\theta \, \frac{h}{2} + M g \cos\theta \, c = M g \cos\theta \, \frac{b}{2}, \tag{10.53}$$

where use has been made of Equations (10.48) and (10.51). The previous equation yields

$$c = \frac{1}{2}\,(b - \mu\, h). \tag{10.54}$$

Now, the physical constraints on c are $0 \le c \le b$. However, if the height of the block, h, exceeds the critical value b/μ then c becomes negative, implying that the equilibrium that we have just been studying is impossible. Hence, we deduce that if $h < b/\mu$ then, as the inclination of the slope is gradually increased, the block will eventually slide down the slope. However, if $h > b/\mu$ then the block will eventually tumble down the slope.

To try to make the arguments of this section more exact, let x measure distance up the slope from the forward corner of the block, and let $r(x)$ be the normal reaction per unit length exerted on the block by the slope. It follows that

$$\int_0^b r(x)\,dx = R, \tag{10.55}$$

and

$$\int_0^b x\,r(x)\,dx = c\,R. \tag{10.56}$$

In other words, the resultant reaction provides the same net force and net torque as the distributed reactions. We deduce that

$$c = \int_0^b x\,r(x)\,dx \Big/ \int_0^b r(x)\,dx. \tag{10.57}$$

However, $r(x) \ge 0$ for $0 \le x \le b$ (i.e., the normal reaction per unit length cannot be negative). Hence, the previous equation cannot be satisfied if $c < 0$. Likewise

$$b - c = \int_0^b (b - x)\,r(x)\,dx \Big/ \int_0^b r(x)\,dx, \tag{10.58}$$

which cannot be satisfied if $c > b$.

10.8 EXERCISES

10.1 Consider a system consisting of N point particles. Let $\mathbf{r}_i$ be the position vector of the ith particle and let $\mathbf{F}_i$ be the external force acting on this particle. Any internal forces are assumed to be central in nature. The resultant force and torque (about the origin) acting on the system are

$$\mathbf{F} = \sum_{i=1,N} \mathbf{F}_i,$$

$$\boldsymbol{\tau} = \sum_{i=1,N} \mathbf{r}_i \times \mathbf{F}_i,$$

respectively. A *point of action* of the resultant force is defined as a point whose position vector, $\mathbf{r}$, satisfies

$$\mathbf{r} \times \mathbf{F} = \boldsymbol{\tau}.$$

Demonstrate that there are an infinite number of possible points of action lying on the straight-line

$$\mathbf{r} = \frac{\mathbf{F} \times \boldsymbol{\tau}}{F^2} + \lambda \frac{\mathbf{F}}{F},$$

where λ is arbitrary. This straight-line is known as the *line of action* of the resultant force.

10.2 Consider an isolated system consisting of two extended bodies (which can, of course, be modeled as collections of point particles), A and B. Let $\mathbf{F}_A$ be the resultant force acting on A due to B and let $\mathbf{F}_B$ be the resultant force acting on B due to A. Demonstrate that $\mathbf{F}_B = -\mathbf{F}_A$, and that both forces have the same line of action.

10.3 Show that an initially stationary rigid body acted upon by two forces can only remain stationary if the forces have the same line of action.

10.4 An extended body is acted upon by two resultant forces, $\mathbf{F}_1$ and $\mathbf{F}_2$. Show that these forces can be only replaced by a single equivalent force, $\mathbf{F} = \mathbf{F}_1 + \mathbf{F}_2$, provided:

(a) $\mathbf{F}_1$ and $\mathbf{F}_2$ are parallel (or antiparallel). In this case, the line of action of $\mathbf{F}$ is parallel to those of $\mathbf{F}_1$ and $\mathbf{F}_2$.

(b) $\mathbf{F}_1$ and $\mathbf{F}_2$ are not parallel (or antiparallel), but their lines of action cross at a point. In this case, the line of action of $\mathbf{F}$ passes through the crossing point.

10.5 Deduce that if an isolated system consists of three extended bodies, A, B, and C, where $\mathbf{F}_A$ is the resultant force acting on A (due to B and C), $\mathbf{F}_B$ is the resultant force acting on B, and $\mathbf{F}_C$ is the resultant force acting on C, then $\mathbf{F}_A + \mathbf{F}_B + \mathbf{F}_C = \mathbf{0}$ and the forces either all have parallel lines of action or have lines of action that cross at a common point.

10.6 Show that an initially stationary rigid body acted upon by three forces can only remain stationary if the forces either have parallel lines of action or have lines of action that cross at a common point.

10.7 Suppose that two uniform rods (of negligible thickness) are welded together at right-angles, as shown in the Figure 10.7. Let the first rod be of mass $m_1 = 5.2\,\mathrm{kg}$ and length $l_1 = 1.3\,\mathrm{m}$. Let the second rod be of mass $m_2 = 3.4\,\mathrm{kg}$ and length $l_2 = 0.7\,\mathrm{m}$. Suppose that the system is suspended from a pivot point located at the free end of the first rod, and then allowed to reach a stable equilibrium state. What angle θ does the first rod subtend with the downward vertical in this state? [Ans: 8.65°.]

10.8 A uniform rod of mass $m = 15\,\mathrm{kg}$ and length $l = 3\,\mathrm{m}$ is supported in a horizontal position by a pin and a cable, as shown in Figure 10.8. Masses $m_1 = 36\,\mathrm{kg}$ and $m_2 = 24\,\mathrm{kg}$ are suspended from the rod at positions $l_1 = 0.5\,\mathrm{m}$ and $l_2 = 2.3\,\mathrm{m}$. The angle θ is 40°. What is the tension T in the cable? [Ans: 486.84 N.]

10.9 A uniform ladder of mass $m = 40\,\mathrm{kg}$ and length $l = 10\,\mathrm{m}$ is leaned against a smooth vertical wall. A person of mass $M = 80\,\mathrm{kg}$ stands on the ladder a distance $x = 7\,\mathrm{m}$ from the bottom, as measured along the ladder. The foot of the ladder is $d = 1.2\,\mathrm{m}$ from the bottom of the wall.

(a) What is the force exerted by the wall on the ladder? [Ans: 90.09 N.]

(b) What is the normal force exerted by the floor on the ladder? [Ans: 1177.2 N.]

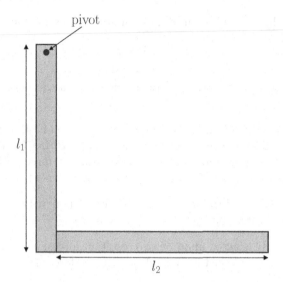

Figure 10.7 Figure for Exercise 10.8

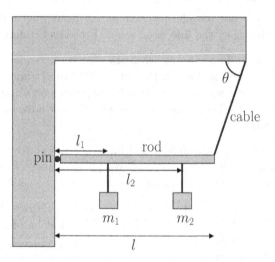

Figure 10.8 Figure for Exercise 10.7

10.10 Determine the smallest angle, θ, with respect to the horizontal for the equilibrium of a homogeneous ladder of length l leaning against a wall. The coefficient of friction for all surfaces is μ. [Ans: $\tan^{-1}[(1 - \mu^2)/(2\mu)]$ for $\mu < 1$, 0 for $\mu > 1$.]

10.11 Consider a light ladder leaning against a vertical wall, as shown in Figure 10.4. Let l be the length of the ladder, and θ its angle of inclination to the horizontal. Suppose that the coefficient of friction between the ladder and the wall, μ, is the same as the coefficient of friction between the ladder and the ground. What is the farthest distance, x, that a person can climb along the ladder (measured from the bottom of the ladder) before it slips? [Ans: $x/l = [\mu/(1 + \mu^2)]\,(\tan\theta + \mu)$.]

10.12 A truck of mass $M = 5000\,\text{kg}$ is crossing a uniform horizontal bridge of mass $m = 1000\,\text{kg}$ and length $l = 100\,\text{m}$. The bridge is supported at its two end-points. What are the reactions at the left and right supports when the truck is one third of the way (from the left) across the bridge? [Ans: 3.76×10^4 N and 2.13×10^4 N.]

10.13 A uniform horizontal rod of mass $m = 15$ kg is attached to a vertical wall at one end, and is supported, from below, by a light rigid strut at the other. The strut is attached to the free end of the rod at one end, and to the wall at the other, and subtends an angle of $\theta = 30°$ with the rod.

(a) What is the horizontal force acting on the strut at the point where it is attached to the rod? (Forces directed away from the wall are positive.) [Ans: -127.44 N.]

(b) What is the vertical force acting on the strut at the point where it is attached to the rod? (Upward forces are positive.) [Ans: -73.58 N.]

(c) What is the horizontal force acting on the strut at the point where it is attached to the wall? [Ans: $+127.44$ N.]

(d) What is the vertical force acting on the strut at the point where it is attached to the wall? [Ans: $+73.58$ N.]

10.14 A uniform rectangular packing case of base-length b and height h lies on a rough horizontal surface. Let M be the mass of the case, and let μ be the coefficient of friction between the case and the surface. A forward horizontal force F acts on the case at its forward top edge. Suppose that the force is gradually increased. What condition must be satisfied if the case is to eventually slide without tipping? [Ans: $h < b/(2\mu)$.]

CHAPTER 11

Oscillatory Motion

11.1   INTRODUCTION

We have seen previously (for instance, in Section 10.3) that if a dynamical system is perturbed from a stable equilibrium state then it experiences a restoring force that acts to return it to that state. In many cases of interest, the magnitude of the restoring force is directly proportional to the displacement from the equilibrium state. In this chapter, we shall investigate the motion of dynamical systems subject to such a force. As we shall see, such systems exhibit **oscillatory** motion.

11.2   SIMPLE HARMONIC MOTION

Let us reexamine the problem of a mass on the end of a spring. (See Section 5.6.) Consider a mass, m, that slides over a horizontal frictionless surface. Suppose that the mass is attached to a light horizontal spring whose other end is anchored in an immovable wall. See Figure 5.8. Let x be the extension of the spring; that is, the difference between the spring's actual length and its unstretched length. Obviously, x can also be used as a coordinate to determine the horizontal displacement of the mass.

The equilibrium state of the system corresponds to the situation in which the mass is at rest, and the spring is unextended (i.e., $x = \dot{x} = 0$). In this state, zero net force acts on the mass, so there is no reason for it to start to move. If the system is perturbed from its equilibrium state (i.e, if the mass is displaced, so that the spring becomes extended) then the mass experiences a restoring force given by Hooke's law:

$$f = -k\,x. \qquad (11.1)$$

Here, $k > 0$ is the *force constant* of the spring. The negative sign indicates that f is indeed a restoring force. Note that the magnitude of the restoring force is directly proportional to the displacement of the system from equilibrium (i.e., $|f| \propto x$). Of course, Hooke's law only holds for relatively small spring extensions. Hence, the displacement from equilibrium cannot be made too large. The motion of this particular system is representative of the motion of a wide range of mechanical systems when they are slightly disturbed from a stable equilibrium state.

Newton's second law gives following equation of motion for the system:

$$m\,\ddot{x} = -k\,x. \qquad (11.2)$$

DOI: 10.1201/9781003198642-11

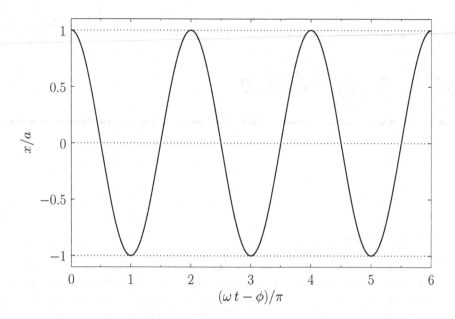

Figure 11.1 Simple harmonic motion. (Reproduced from Fitzpatrick 2019. Courtesy of Taylor & Francis.)

The previous differential equation is known as the *simple harmonic equation*, and its solution has been known for centuries. In fact, the solution is

$$x = a \, \cos(\omega \, t - \phi), \tag{11.3}$$

where a, ω, and ϕ are constants. We can demonstrate that Equation (11.3) is indeed a solution of Equation (11.2) by direct substitution. Substituting Equation (11.3) into Equation (11.2), and recalling from calculus that $d(\cos \theta)/d\theta = -\sin \theta$ and $d(\sin \theta)/d\theta = \cos \theta$, we obtain

$$-m \, \omega^2 \, a \, \cos(\omega \, t - \phi) = -k \, a \, \cos(\omega \, t - \phi). \tag{11.4}$$

It follows that Equation (11.3) is the correct solution provided

$$\omega = \sqrt{\frac{k}{m}}. \tag{11.5}$$

Figure 11.1 shows a graph of x versus t obtained from Equation (11.3). The type of motion shown in the figure is know as *simple harmonic motion*. It can be seen that the displacement, x, oscillates between $x = -a$ and $x = +a$. Here, a is termed the *amplitude* of the oscillation. Moreover, the motion is periodic in time (i.e., it repeats exactly after a certain time period has elapsed). In fact, the period of the motion is

$$T = \frac{2\pi}{\omega}. \tag{11.6}$$

To be more exact, $x(t+T) = x(t)$ for all t. This result is easily obtained from Equation (11.3) by noting that $\cos \theta$ is a periodic function of θ with period 2π [i.e., $\cos(\theta + 2\pi) = \cos \theta$ for all θ]. The *frequency* of the motion (i.e., the number of oscillations completed per second) is

$$f = \frac{1}{T} = \frac{\omega}{2\pi}. \tag{11.7}$$

Table 11.1 Simple Harmonic Motion.
(Reproduced from Fitzpatrick 2019. Courtesy
of Taylor & Francis.)

$\omega t - \phi$	0	$\pi/2$	π	$3\pi/2$
x	$+a$	0	$-a$	0
$\dot{x}$	0	$-\omega a$	0	$+\omega a$
$\ddot{x}$	$-\omega^2 a$	0	$+\omega^2 a$	0

It can be seen that ω is the motion's *angular frequency* (i.e., the frequency, f, converted into radians per second). Finally, the *phase angle*, ϕ, determines the times at which the oscillation attains its maximum amplitude, $x = a$; in fact,

$$t_{\max} = \left(n + \frac{\phi}{2\pi}\right) T, \tag{11.8}$$

where n is an arbitrary integer.

Table 11.1 lists the displacement, velocity, and acceleration of the mass at various phases of the simple harmonic cycle. The information contained in this table can easily be derived from the simple harmonic solution, Equation (11.3). Note that all of the non-zero values shown in this table represent either the maximum or the minimum value taken by the quantity in question during the oscillation cycle.

We have seen that when a mass on a spring is disturbed it executes simple harmonic motion about its equilibrium state. In physical terms, if the initial displacement is positive $(x > 0)$ then the restoring force overcompensates, and sends the system past the equilibrium state $(x = 0)$ to negative displacement states $(x < 0)$. The restoring force again overcompensates, and sends the system back through $x = 0$ to positive displacement states. The motion then repeats itself ad infinitum. The angular frequency of the oscillation is determined by the spring stiffness, k, and the system inertia, m, via Equation (11.5). In contrast, the amplitude and phase angle of the oscillation are determined by the initial conditions.

Suppose that the instantaneous displacement and velocity of the mass at $t = 0$ are x_0 and v_0, respectively. It follows from Equation (11.3) that

$$x_0 \equiv x(t = 0) = a \cos \phi, \tag{11.9}$$

$$v_0 \equiv \dot{x}(t = 0) = a\omega \sin \phi. \tag{11.10}$$

Here, use has been made of the well-known identities $\cos(-\theta) = \cos\theta$ and $\sin(-\theta) = -\sin\theta$. Hence, we obtain

$$a = \sqrt{x_0^2 + (v_0/\omega)^2}, \tag{11.11}$$

and

$$\phi = \tan^{-1}\left(\frac{v_0}{\omega x_0}\right), \tag{11.12}$$

because $\sin^2\theta + \cos^2\theta = 1$ and $\tan\theta = \sin\theta/\cos\theta$.

The kinetic energy of the system is written

$$K = \frac{1}{2}m\dot{x}^2 = \frac{1}{2}ma^2\omega^2\sin^2(\omega t - \phi). \tag{11.13}$$

Recall, from Section 5.6, that the potential energy takes the form

$$U = \frac{1}{2}kx^2 = \frac{1}{2}ka^2\cos^2(\omega t - \phi). \tag{11.14}$$

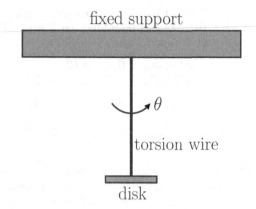

Figure 11.2 A torsion pendulum.

Hence, the total energy can be written

$$E = K + U = \frac{1}{2} k a^2,$$
(11.15)

because $m \omega^2 = k$ and $\sin^2 \theta + \cos^2 \theta = 1$. Note that the total energy is a constant of the motion, as is to be expected for an isolated system subject to a conservative force. Moreover, the energy is proportional to the square of the amplitude of the motion. It is clear, from the previous expressions, that simple harmonic motion is characterized by a constant backward and forward flow of energy between kinetic and potential components. The kinetic energy attains its maximum value, and the potential energy attains it minimum value, when the displacement is zero (i.e., when $x = 0$). Likewise, the potential energy attains its maximum value, and the kinetic energy attains its minimum value, when the displacement is maximal (i.e., when $x = \pm a$). Note that the minimum value of K is zero, because the system is instantaneously at rest when the displacement is maximal.

11.3 TORSION PENDULUM

Consider a horizontal disk that is suspended from a vertical torsion wire attached to its center. See Figure 11.2. This setup is known as a *torsion pendulum*. A torsion wire is essentially inextensible, but is free to twist about its axis. Of course, as the wire twists it also causes the disk attached to it to rotate in a horizontal plane. Let θ be the angle of rotation of the disk, and let $\theta = 0$ correspond to the case in which the wire is untwisted.

Any twisting of the wire is inevitably associated with mechanical deformation. The wire resists such deformation by developing a restoring torque, τ, that acts to restore the wire to its untwisted state. For relatively small angles of twist, the magnitude of this torque is directly proportional to the twist angle. Hence, we can write

$$\tau = -k \theta,$$
(11.16)

where $k > 0$ is the *torque constant* of the wire. The previous equation is essentially a torsional equivalent to Hooke's law. The rotational equation of motion of the system is written

$$I \ddot{\theta} = \tau,$$
(11.17)

where I is the moment of inertia of the disk (about a perpendicular axis passing through its center). The moment of inertia of the wire is assumed to be negligible. Combining the previous two equations, we obtain

$$I\ddot{\theta} = -k\,\theta. \tag{11.18}$$

Equation (11.18) is clearly a simple harmonic equation [cf., Equation (11.2)]. Hence, we can immediately write the standard solution [cf., Equation (11.3)]

$$\theta = a\,\cos(\omega\,t - \phi), \tag{11.19}$$

where [cf., Equation (11.5)]

$$\omega = \sqrt{\frac{k}{I}}. \tag{11.20}$$

We conclude that if a torsion pendulum is perturbed from its equilibrium state then it executes torsional oscillations about that state at a fixed angular frequency, ω, that only depends on the torque constant of the wire and the moment of inertia of the disk. Note, in particular, that the frequency is independent of the amplitude of the oscillation [provided that θ remains small enough that Equation (11.16) still applies]. Torsion pendulums are often used for time-keeping purposes. For instance, the balance wheel in a mechanical wristwatch is a torsion pendulum in which the restoring torque is provided by a coiled spring.

11.4   SIMPLE PENDULUM

Consider a compact mass, m, suspended from a light inextensible string of length l, such that the mass is free to swing from side to side in a vertical plane, as shown in Figure 11.3. This setup is known as a *simple pendulum*. Let θ be the angle subtended between the string and the downward vertical. Obviously, the equilibrium state of the simple pendulum corresponds to the situation in which the mass is stationary and is located directly below the suspension point (i.e., $\theta = \dot{\theta} = 0$). The angular equation of motion of the pendulum is simply

$$I\ddot{\theta} = \tau, \tag{11.21}$$

where I is the moment of inertia of the mass, and τ is the torque acting on the system. For the case in hand, given that the mass is essentially a point particle, and is situated a distance l from the axis of rotation (i.e., the suspension point), it is easily seen that $I = m\,l^2$.

The two forces acting on the mass are the downward gravitational force, $m\,g$, and the tension, T, in the string. Note, however, that the tension makes no contribution to the torque, because its line of action clearly passes through the axis of rotation. From simple trigonometry, the line of action of the gravitational force passes a distance $l\,\sin\theta$ from the axis of rotation. Hence, the magnitude of the gravitational torque is $m\,g\,l\,\sin\theta$. Moreover, the gravitational torque is a restoring torque; that is, if the mass is displaced slightly from its equilibrium state then the gravitational torque clearly acts to push the mass back toward that state. Thus, we can write

$$\tau = -m\,g\,l\,\sin\theta. \tag{11.22}$$

Combining the previous two equations, we obtain the following angular equation of motion of the pendulum:

$$l\ddot{\theta} = -g\,\sin\theta. \tag{11.23}$$

Unfortunately, this is not the simple harmonic equation. Indeed, the previous equation does not possess a closed solution that can be expressed in terms of simple functions.

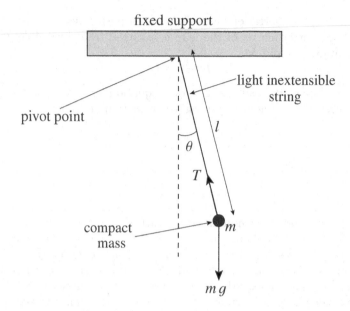

Figure 11.3 A simple pendulum. (Reproduced from Fitzpatrick 2019. Courtesy of Taylor & Francis.)

Suppose that we restrict our attention to relatively small deviations from the equilibrium state. In other words, suppose that the angle θ is constrained to take fairly small values. We know, from trigonometry, that for $|\theta|$ less than about 6° it is a good approximation to write

$$\sin\theta \simeq \theta. \tag{11.24}$$

Hence, in the small-angle limit, Equation (11.23) reduces to

$$l\,\ddot{\theta} = -g\,\theta, \tag{11.25}$$

which is in the familiar form of a simple harmonic equation. Comparing with our original simple harmonic equation, Equation (11.2), and its solution, we conclude that the angular frequency of small-amplitude oscillations of a simple pendulum is given by

$$\omega = \sqrt{\frac{g}{l}}. \tag{11.26}$$

In this case, the pendulum frequency is dependent only on the length of the pendulum and the local gravitational acceleration, and is independent of the mass of the pendulum and the amplitude of the pendulum swings (provided that $\sin\theta \simeq \theta$ remains a good approximation).

In reality, the period of a simple pendulum is amplitude dependent. In fact, if θ_0 is the amplitude of the motion, T the period of the motion, and T_0 the small-amplitude period, then

$$\frac{T}{T_0} = 1 + \frac{\theta_0^2}{16} + \frac{11\,\theta_0^4}{3072} + \cdots . \tag{11.27}$$

So, the period of the pendulum increases with amplitude. It is clear that a simple pendulum can only be used as an accurate time keeping device if the amplitude of the oscillation is kept fixed. This goal is achieved by means of an *escapement mechanism*. Simple pendulums can also be used to measure local variations in gravitational acceleration on the surface of the Earth.

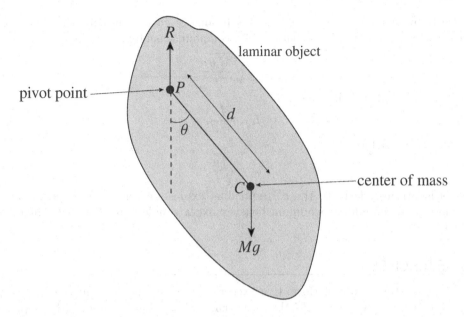

Figure 11.4 A compound pendulum. (Reproduced from Fitzpatrick 2019. Courtesy of Taylor & Francis.)

11.5 COMPOUND PENDULUM

Consider an extended laminar (i.e., flat) object of mass M with a hole drilled though it. Suppose that the object is suspended from a fixed peg, which passes through the hole, such that it is free to swing from side to side in a vertical plane, as shown in Figure 11.4. This setup is known as a *compound pendulum*.

Let P be the pivot point, and let C be the object's center of mass, which is located a distance d from the pivot. Let θ be the angle subtended between the downward vertical (which passes through point P) and the line PC. The equilibrium state of the compound pendulum corresponds to the case in which the center of mass lies vertically below the pivot point; that is, $\theta = 0$. See Section 10.3. The angular equation of motion of the pendulum is simply

$$I\ddot{\theta} = \tau, \tag{11.28}$$

where I is the moment of inertia of the object about the pivot point and τ is the torque. Using similar arguments to those employed for the case of the simple pendulum (recalling that the resultant weight of the pendulum acts at its center of mass), we can write

$$\tau = -M\,g\,d\,\sin\theta. \tag{11.29}$$

Note that the reaction, R, at the peg does not contribute to the torque, because its line of action passes through the pivot point. Combining the previous two equations, we obtain the following angular equation of motion of the pendulum:

$$I\ddot{\theta} = -M\,g\,d\,\sin\theta. \tag{11.30}$$

Finally, adopting the small-angle approximation, $\sin\theta \simeq \theta$, we arrive at the simple harmonic equation:

$$I\ddot{\theta} = -M\,g\,d\,\theta. \tag{11.31}$$

It is clear, by analogy with our previous solutions of such equations, that the angular frequency of small-amplitude oscillations of a compound pendulum is given by

$$\omega = \sqrt{\frac{M g d}{I}}. \tag{11.32}$$

It is helpful to define the length

$$L = \frac{I}{M d}. \tag{11.33}$$

Equation (11.32) reduces to

$$\omega = \sqrt{\frac{g}{L}}, \tag{11.34}$$

which is identical in form to the corresponding expression for a simple pendulum. We conclude that a compound pendulum behaves like a simple pendulum with an *effective length L*.

11.6 EXERCISES

11.1 A mass stands on a platform that executes simple harmonic oscillation in a vertical direction at a frequency of 5 Hz. Show that the mass loses contact with the platform when the displacement exceeds 10^{-2} m. [From Pain 1999]

11.2 Two light springs have spring constants k_1 and k_2, respectively, and are used in a vertical orientation to support an object of mass m. Show that the angular frequency of small amplitude oscillations about the equilibrium state is $[(k_1 + k_2)/m]^{1/2}$ if the springs are connected in parallel, and $[k_1 k_2/(k_1 + k_2) m]^{1/2}$ if the springs are connected in series. [From Fitzpatrick 2019.]

11.3 A body of uniform cross-sectional area A and mass density ρ floats in a liquid of density ρ_0 (where $\rho < \rho_0$), and at equilibrium displaces a volume V. Making use of *Archimedes' principle* (that the buoyancy force acting on a partially submerged body is equal to the weight of the displaced liquid), show that the period of small-amplitude oscillations about the equilibrium position is

$$T = 2\pi \sqrt{\frac{V}{g A}}.$$

[From Fitzpatrick 2019.]

11.4 A piston in a stream engine executes simple harmonic motion. The maximum displacement of the piston from its center-line is ± 7 cm, The mass of the piston is 4 kg. Finally, the engine is running at 4000 rev/min.

(a) What is the maximum velocity of the piston? [Ans: 29.32 m/s.]
(b) What is the maximum acceleration of the piston? [Ans: 1.228×10^4 m/s^2.]

11.5 A block attached to a spring executes simple harmonic motion in a horizontal direction with an amplitude of 0.25 m. At a point 0.15 m away from the equilibrium position, the velocity of the block is 0.75 m/s. What is the period of oscillation of the block? [Ans: 1.676 s.]

11.6 A block of mass $m = 3$ kg is attached to two springs, as shown in Figure 11.5 and slides over a horizontal frictionless surface. Given that the force constants of the two springs are $k_1 = 1200$ N/m and $k_2 = 400$ N/m, find the period of oscillation of the system. [Ans: 0.6283 s.]

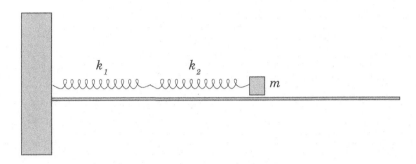

Figure 11.5 Figure for Exercise 11.6

11.7 A block of mass $m = 4\,\text{kg}$ is attached to a spring, and undergoes simple harmonic motion with a period of $T = 0.35\,\text{s}$. The total energy of the system is $E = 2.5\,\text{J}$.

(a) What is the force constant of the spring? [Ans: $1289.1\,\text{N/m}$.]

(b) What is the amplitude of the motion? [Ans: $0.06228\,\text{m}$.]

11.8 Having landed on a newly discovered planet, an astronaut sets up a simple pendulum of length $0.6\,\text{m}$, and finds that it makes 51 complete oscillations in 1 minute. The amplitude of the oscillations is small compared to the length of the pendulum. What is the surface gravitational acceleration on the planet? [Ans: $17.11\,\text{m/s}^2$.]

11.9 A uniform disk of radius $r = 0.8\,\text{m}$ and mass $M = 3\,\text{kg}$ is freely suspended from a horizontal pivot located a radial distance $d = 0.25\,\text{m}$ from its center, such that the disk can oscillate in a vertical plane. Find the angular frequency of small amplitude oscillations of the disk. [Ans: $2.532\,\text{rad/s}$.]

11.10 A compound pendulum consists of a uniform circular disk of radius r that is free to turn about a horizontal axis perpendicular to its plane. Find the position of the axis for which the periodic time is a minimum. [Ans: $r/\sqrt{2}$ from the center of the disk.]

11.11 A laminar object of mass M has a moment of inertia I_0 about a perpendicular axis passing through its center of mass. Suppose that the object is converted into a compound pendulum by suspending it about a horizontal axis perpendicular to its plane. Show that the minimum effective length of the pendulum occurs when the distance of the suspension point from the center of gravity is equal to the radius of gyration, $k = (I_0/M)^{1/2}$.

11.12 A compound pendulum consists of a uniform bar of length l that pivots in a vertical plane about one of its ends. Show that the pendulum has the same period of oscillation as a simple pendulum of length $(2/3)\,l$. [From Fitzpatrick 2019.]

11.13 A particle of mass m is attached to a light wire that is stretched tightly between two fixed points with a tension T. The particle is located a distance a from one end of the wire, and a distance b from the other.

(a) Show that the period of small transverse oscillations of the particle is

$$2\pi\sqrt{\frac{m}{T}\frac{a\,b}{a+b}}.$$

(b) Show that for a wire of fixed length the period is longest when the particle is attached to the middle point.

CHAPTER 12

Rotating Reference Frames

12.1 INTRODUCTION

As we saw in Section 4.12, Newton's second law of motion is only valid in inertial frames of reference. Unfortunately, we are sometimes forced to observe motion in non-inertial reference frames. For instance, it is most convenient for us to observe the motions of the objects in our immediate vicinity in a reference frame that is fixed relative to the surface of the Earth. Such a frame is non-inertial in nature, because it accelerates with respect to a standard inertial frame due to the Earth's diurnal rotation about its axis. (Note that the accelerations of this frame due to the Earth's orbital motion about the Sun, or the Sun's orbital motion about the galactic center, etcetera, are negligible compared to the acceleration due to the Earth's diurnal rotation.) Let us now investigate motion in a **rotating reference frame**.

12.2 ROTATING REFERENCE FRAMES

Suppose that a given object has position vector $\mathbf{r}$ in some non-rotating inertial reference frame. Let us observe the motion of this object in a non-inertial reference frame that rotates with constant angular velocity $\boldsymbol{\Omega}$ about an axis passing through the origin of the inertial frame. Suppose, first of all, that our object appears stationary in the rotating reference frame. Hence, in the non-rotating frame, the object's position vector, $\mathbf{r}$, will appear to rotate about the origin with angular velocity $\boldsymbol{\Omega}$. It follows, from Equation (8.12), that in the non-rotating reference frame

$$\frac{d\mathbf{r}}{dt} = \boldsymbol{\Omega} \times \mathbf{r}. \tag{12.1}$$

Suppose, now, that our object appears to move in the rotating reference frame with instantaneous velocity $\mathbf{v}'$. It is fairly obvious that the appropriate generalization of the previous equation is simply

$$\frac{d\mathbf{r}}{dt} = \mathbf{v}' + \boldsymbol{\Omega} \times \mathbf{r}. \tag{12.2}$$

Let and d/dt and d/dt' denote apparent time derivatives in the non-rotating and rotating frames of reference, respectively. Because an object that is stationary in the rotating reference frame appears to move in the non-rotating frame, it is clear that $d/dt \neq d/dt'$. Writing the apparent velocity, $\mathbf{v}'$, of our object in the rotating reference frame as $d\mathbf{r}/dt'$, the previous equation takes the form

$$\frac{d\mathbf{r}}{dt} = \frac{d\mathbf{r}}{dt'} + \boldsymbol{\Omega} \times \mathbf{r}, \tag{12.3}$$

DOI: 10.1201/9781003198642-12

or

$$\frac{d}{dt} = \frac{d}{dt'} + \boldsymbol{\Omega} \times,$$ (12.4)

because **r** is a general position vector. Equation (12.4) expresses the relationship between apparent time derivatives in the non-rotating and rotating reference frames.

Operating on the general position vector **r** with the time derivative (12.4), we get

$$\mathbf{v} = \mathbf{v}' + \boldsymbol{\Omega} \times \mathbf{r}.$$ (12.5)

This equation relates the apparent velocity, $\mathbf{v} = d\mathbf{r}/dt$, of an object with position vector **r** in the non-rotating reference frame to its apparent velocity, $\mathbf{v}' = d\mathbf{r}/dt'$, in the rotating reference frame.

Operating twice on the position vector **r** with the time derivative (12.4), we obtain

$$\mathbf{a} = \left(\frac{d}{dt'} + \boldsymbol{\Omega} \times\right)(\mathbf{v}' + \boldsymbol{\Omega} \times \mathbf{r}),$$ (12.6)

or

$$\mathbf{a} = \mathbf{a}' + \boldsymbol{\Omega} \times (\boldsymbol{\Omega} \times \mathbf{r}) + 2\,\boldsymbol{\Omega} \times \mathbf{v}'.$$ (12.7)

This equation relates the apparent acceleration, $\mathbf{a} = d\mathbf{v}/dt = d^2\mathbf{r}/dt^2$, of an object with position vector **r** in the non-rotating reference frame to its apparent acceleration, $\mathbf{a}' = d\mathbf{v}'/dt' = d^2\mathbf{r}/dt'^2$, in the rotating reference frame.

Applying Newton's second law of motion in the inertial (i.e., non-rotating) reference frame, we obtain

$$m\,\mathbf{a} = \mathbf{f}.$$ (12.8)

Here, m is the mass of our object, and **f** is the (non-fictitious) force acting on it. Note that these quantities are the same in both reference frames. Making use of Equation (12.7), the apparent equation of motion of our object in the rotating reference frame takes the form

$$m\,\mathbf{a}' = \mathbf{f} - m\,\boldsymbol{\Omega} \times (\boldsymbol{\Omega} \times \mathbf{r}) - 2\,m\,\boldsymbol{\Omega} \times \mathbf{v}'.$$ (12.9)

The last two terms on the right-hand side of the previous equation are so-called *fictitious forces*. Such forces are always needed to account for motion observed in non-inertial reference frames. Note that fictitious forces can always be distinguished from non-fictitious forces in Newtonian dynamics because the former have no associated reactions. Let us now investigate the two fictitious forces appearing in Equation (12.9).

12.3 CENTRIFUGAL ACCELERATION

Let our non-rotating inertial frame be one whose origin lies at the center of the Earth, and let our rotating frame be one whose origin is fixed with respect to some point, of latitude λ, on the Earth's surface. See Figure 12.1. The latter reference frame thus rotates with respect to the former (about an axis passing through the Earth's center) with an angular velocity vector, $\boldsymbol{\Omega}$, that points from the center of the Earth toward its north pole, and is of magnitude

$$\Omega = \frac{2\pi}{23^{\mathrm{h}}\,56^{\mathrm{m}}\,04^{\mathrm{s}}} = 7.2921 \times 10^{-5}\,\mathrm{rad/s}$$ (12.10)

(Yoder 1995). Here, $23^{\mathrm{h}}\,56^{\mathrm{m}}\,04^{\mathrm{s}}$ is the length of a so-called *sidereal day*, and is the period of the Earth's rotation relative to distant stars (as opposed to the Sun).

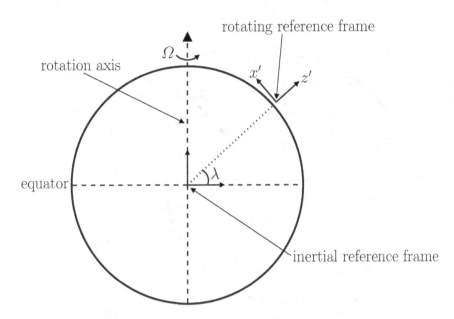

Figure 12.1 Inertial and non-inertial reference frames. (Reproduced from Fitzpatrick 2012. Courtesy of Cambridge University Press.)

Consider an object that appears stationary in our rotating reference frame; in other words, an object that is stationary with respect to the Earth's surface. According to Equation (12.9), the object's apparent equation of motion in the rotating frame takes the form

$$m \, \mathbf{a}' = \mathbf{f} - m \, \mathbf{\Omega} \times (\mathbf{\Omega} \times \mathbf{r}). \tag{12.11}$$

Let the non-fictitious force acting on our object be the force of gravity, $\mathbf{f} = m \, \mathbf{g}$. Here, the local gravitational acceleration, $\mathbf{g}$, points directly toward the center of the Earth (assuming, for the sake of simplicity, that the Earth is spherical). It follows, from the previous analysis, that the apparent gravitational acceleration in the rotating frame is written

$$\mathbf{g}' = \mathbf{g} - \mathbf{\Omega} \times (\mathbf{\Omega} \times \mathbf{R}), \tag{12.12}$$

where $\mathbf{R}$ is the displacement vector of the origin of the rotating frame (which lies on the Earth's surface) with respect to the center of the Earth. Here, we are assuming that our object is situated relatively close to the Earth's surface (i.e., $\mathbf{r} \simeq \mathbf{R}$).

It can be seen, from Equation (12.12), that the apparent gravitational acceleration of a stationary object close to the Earth's surface has two components. First, the true gravitational acceleration, $\mathbf{g}$, of magnitude $g \simeq 9.81 \, \mathrm{m/s^2}$, that always points directly toward the center of the Earth. Second, the so-called *centrifugal acceleration*, $-\mathbf{\Omega} \times (\mathbf{\Omega} \times \mathbf{R})$. This acceleration is normal to the Earth's axis of rotation, and always points directly away from this axis. The magnitude of the centrifugal acceleration is $\Omega^2 \rho = \Omega^2 R \cos \lambda$, where ρ is the perpendicular distance to the Earth's rotation axis, and $R = 6.371 \times 10^6 \, \mathrm{m}$ is the Earth's mean radius (Yoder 1995). See Figure 12.2.

It is convenient to define Cartesian axes in the rotating reference frame such that the z'-axis points vertically upward, and x'- and y'-axes are horizontal, with the x'-axis pointing directly northward and the y'-axis pointing directly westward. See Figure 12.1. The Cartesian components of the Earth's angular velocity are thus

$$\mathbf{\Omega} = \Omega \, (\cos \lambda, \, 0, \, \sin \lambda), \tag{12.13}$$

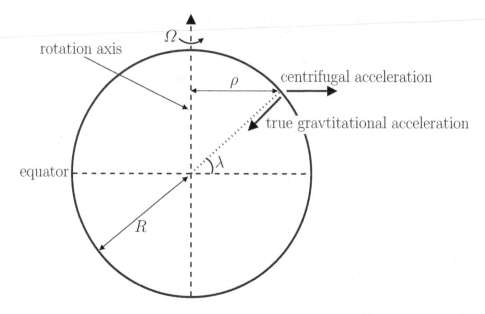

Figure 12.2 Centrifugal acceleration. (Reproduced from Fitzpatrick 2012. Courtesy of Cambridge University Press.)

while the vectors $\mathbf{R}$ and $\mathbf{g}$ are written

$$\mathbf{R} = (0,\, 0,\, R), \qquad (12.14)$$

$$\mathbf{g} = (0,\, 0,\, -g), \qquad (12.15)$$

respectively. It follows that the Cartesian coordinates of the apparent gravitational acceleration, (12.12), are

$$\mathbf{g}' = \left(-\Omega^2\, R\, \cos\lambda\, \sin\lambda,\, 0,\, -g + \Omega^2\, R\, \cos^2\lambda\right). \qquad (12.16)$$

The magnitude of this acceleration is approximately

$$g' \simeq g - \Omega^2\, R\, \cos^2\lambda \simeq 9.81 - 0.034\, \cos^2\lambda \ \ \mathrm{m/s^2}. \qquad (12.17)$$

According to the previous equation, the centrifugal acceleration causes the magnitude of the apparent gravitational acceleration on the Earth's surface to vary by about 0.3%, being largest at the poles, and smallest at the equator. This variation in apparent gravitational acceleration, due (ultimately) to the Earth's rotation, causes the Earth itself to bulge slightly at the equator, which has the effect of further intensifying the variation, because a point on the surface of the Earth at the equator is slightly further away from the Earth's center than a similar point at one of the poles (and, hence, the true gravitational acceleration is slightly weaker in the former case). (See Exercise 15.2.)

Another consequence of centrifugal acceleration is that the apparent gravitational acceleration on the Earth's surface has a horizontal component aligned in the north/south direction. This horizontal component ensures that the apparent gravitational acceleration does not point directly toward the center of the Earth. In other words, a plumb-line on the surface of the Earth does not point vertically downward (i.e., toward the center of the Earth), but is deflected slightly away from a true vertical in the north/south direction. The angular deviation from true vertical can easily be calculated from Equation (12.16):

$$\theta_{\text{dev}} \simeq -\frac{\Omega^2 R}{2g} \sin(2\lambda) \simeq -0.1° \sin(2\lambda).$$ (12.18)

Here, a positive angle denotes a northward deflection, and vice versa. Thus, the deflection is southward in the northern hemisphere (i.e., $\lambda > 0$) and northward in the southern hemisphere (i.e., $\lambda < 0$). The deflection is zero at the poles and at the equator, and reaches its maximum magnitude (which is very small) at middle latitudes.

12.4   CORIOLIS FORCE

We have now accounted for the first fictitious force, $-m\,\mathbf{\Omega}\times(\mathbf{\Omega}\times\mathbf{r})$, in Equation (12.9). Let us now investigate the second, which takes the form $-2\,m\,\mathbf{\Omega}\times\mathbf{v}'$, and is called the *Coriolis force*. Obviously, this force only affects objects that are moving in the rotating reference frame.

Consider a particle of mass m free-falling under gravity in our rotating reference frame. As before, we define Cartesian axes in the rotating frame such that the z'-axis points vertically upward, and the x'- and y'-axes are horizontal, with the x'-axis pointing directly northward, and the y'-axis pointing directly westward. It follows, from Equation (12.9), that the Cartesian equations of motion of the particle in the rotating reference frame take the form:

$$\ddot{x}' = 2\,\Omega\,\sin\lambda\,\dot{y}',$$ (12.19)

$$\ddot{y}' = -2\,\Omega\,\sin\lambda\,\dot{x}' + 2\,\Omega\,\cos\lambda\,\dot{z}',$$ (12.20)

$$\ddot{z}' = -g - 2\,\Omega\,\cos\lambda\,\dot{y}'.$$ (12.21)

Here, g is the local acceleration due to gravity. In the previous three equations, we have neglected the centrifugal acceleration, for the sake of simplicity. This is reasonable, because the only effect of the centrifugal acceleration is to slightly modify the magnitude and direction of the local gravitational acceleration. We have also neglected air resistance, which is less reasonable.

Consider a particle that is dropped (at $t = 0$) from rest a height h above the Earth's surface. The following solution method exploits the fact that the Coriolis force is much smaller in magnitude than the force of gravity. Hence, Ω can be treated as a small parameter. To lowest order (i.e., neglecting Ω), the particle's vertical motion satisfies $\ddot{z}' = -g$, which can be solved, subject to the initial conditions, to give

$$z' \simeq h - \frac{g\,t^2}{2}.$$ (12.22)

Substituting this expression into Equations (12.19) and (12.20), neglecting terms involving Ω^2, and solving subject to the initial conditions, we obtain $x' \simeq 0$, and

$$y' \simeq -g\,\Omega\,\cos\lambda\,\frac{t^3}{3}.$$ (12.23)

In other words, the particle is deflected eastward (i.e., in the negative y'-direction). The particle hits the ground when $t \simeq (2\,h/g)^{1/2}$. Hence, the net eastward deflection of the particle as strikes the ground is

$$d_{\text{east}} \simeq \frac{\Omega}{3}\,\cos\lambda\,\left(\frac{8\,h^3}{g}\right)^{1/2}.$$ (12.24)

Note that this deflection is in the same direction as the Earth's rotation (i.e., west to east), and is greatest at the equator, and zero at the poles. A particle dropped from a height of 100 m at the equator is deflected by about 2.2 cm.

Consider a particle launched horizontally with some fairly large velocity

$$\mathbf{V} = V_0 \left(\cos\theta, -\sin\theta, 0 \right). \tag{12.25}$$

Here, θ is the compass bearing of the velocity vector (so, north is $0°$, east is $90°$, etcetera). Neglecting any vertical motion, Equations (12.19) and (12.20) yield

$$\dot{v}_{x'} \simeq -2\,\Omega\,V_0 \sin\lambda \sin\theta, \tag{12.26}$$

$$\dot{v}_{y'} \simeq -2\,\Omega\,V_0 \sin\lambda \cos\theta, \tag{12.27}$$

which can be integrated to give

$$v_{x'} \simeq V_0 \cos\theta - 2\,\Omega\,V_0 \sin\lambda \sin\theta\,t, \tag{12.28}$$

$$v_{y'} \simeq -V_0 \sin\theta - 2\,\Omega\,V_0 \sin\lambda \cos\theta\,t. \tag{12.29}$$

To lowest order in Ω, the previous equations are equivalent to

$$v_{x'} \simeq V_0 \cos(\theta + 2\,\Omega\,\sin\lambda\,t), \tag{12.30}$$

$$v_{y'} \simeq -V_0 \sin(\theta + 2\,\Omega\,\sin\lambda\,t). \tag{12.31}$$

It follows that the Coriolis force causes the compass bearing of the particle's velocity vector to rotate steadily as time progresses. The rotation rate is

$$\frac{d\theta}{dt} \simeq 2\,\Omega\,\sin\lambda. \tag{12.32}$$

Hence, the rotation is clockwise (looking from above) in the northern hemisphere and counter-clockwise in the southern hemisphere. The rotation rate is zero at the equator and greatest at the poles.

The Coriolis force has a significant effect on terrestrial weather patterns. Near equatorial regions, the intense heating of the Earth's surface due to the Sun results in hot air rising. In the northern hemisphere, this causes cooler air to move in a southerly direction toward the equator. The Coriolis force deflects this moving air in a clockwise sense (looking from above), resulting in the *trade winds*, which blow toward the southwest. In the southern hemisphere, the cooler air moves northward, and is deflected by the Coriolis force in a counter-clockwise sense, resulting in trade winds that blow toward the northwest.

Furthermore, as air flows from high to low pressure regions, the Coriolis force deflects the air in a clockwise/counter-clockwise manner in the northern/southern hemisphere, producing *cyclonic rotation*. See Figure 12.3. It follows that cyclonic rotation is counter-clockwise in the northern hemisphere and clockwise in the southern hemisphere. Thus, this is the direction of rotation of tropical storms (e.g., hurricanes, typhoons) in each hemisphere.

12.5 FOUCAULT PENDULUM

Consider a pendulum consisting of a compact mass, m, suspended from a light cable of length l, in such a manner that the pendulum is free to oscillate in any plane whose normal is parallel to the Earth's surface. The mass is subject to three forces. First, the force of gravity, $m\,\mathbf{g}$, which is directed vertically downward (we are again ignoring centrifugal acceleration).

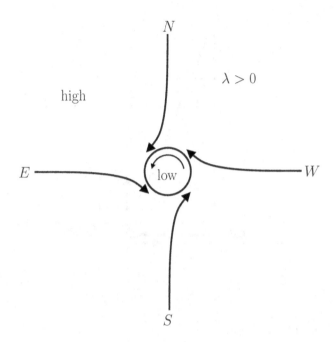

Figure 12.3 A cyclone in the northern hemisphere. (Reproduced from Fitzpatrick 2012. Courtesy of Cambridge University Press.)

Second, the tension, $\mathbf{T}$, in the cable, which is directed upward along the cable. Third, the Coriolis force. It follows that the apparent equation of motion of the mass, in a frame of reference that co-rotates with the Earth, is [see Equation (12.9)]

$$m\,\ddot{\mathbf{r}}' = m\,\mathbf{g} + \mathbf{T} - 2\,m\,\mathbf{\Omega} \times \dot{\mathbf{r}}'. \tag{12.33}$$

Let us define our usual Cartesian coordinates (x', y', z'), and let the origin of our coordinate system correspond to the equilibrium position of the mass. If the pendulum cable is deflected from the downward vertical by a small angle θ then it is easily seen that $x' \sim l\,\theta$, $y' \sim l\,\theta$, and $z' \sim l\,\theta^2$. In other words, the change in height of the mass, z', is negligible compared to its horizontal displacement. Hence, we can write $z' \simeq 0$, provided that $|\theta| \ll 1$. The tension, $\mathbf{T}$, has the vertical component $T\cos\theta \simeq T$, and the horizontal component $\mathbf{T}_{\mathrm{hz}} = -T\sin\theta\,\mathbf{r}'/r' \simeq -T\,\mathbf{r}'/l$, because $\sin\theta \simeq r'/l$. See Figure 12.4. Hence, the Cartesian equations of motion of the mass are written [cf., Equations (12.19)–(12.21)]

$$\ddot{x}' = -\frac{T}{l\,m}\,x' + 2\,\Omega\,\sin\lambda\,\dot{y}', \tag{12.34}$$

$$\ddot{y}' = -\frac{T}{l\,m}\,y' - 2\,\Omega\,\sin\lambda\,\dot{x}', \tag{12.35}$$

$$0 = \frac{T}{m} - g - 2\,\Omega\,\cos\lambda\,\dot{y}'. \tag{12.36}$$

To lowest order in Ω (i.e., neglecting Ω), as well as in $|\theta|$, the final equation, which is just vertical force balance, yields $T \simeq m\,g$. Hence, Equations (12.34) and (12.35) reduce to

$$\ddot{x}' \simeq -\frac{g}{l}\,x' + 2\,\Omega\,\sin\lambda\,\dot{y}', \tag{12.37}$$

$$\ddot{y}' \simeq -\frac{g}{l}\,y' - 2\,\Omega\,\sin\lambda\,\dot{x}'. \tag{12.38}$$

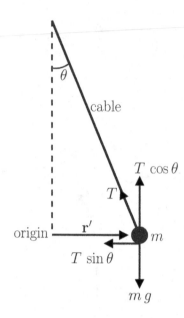

Figure 12.4 The Foucault pendulum

Let

$$s = x' + i\,y'. \tag{12.39}$$

Equations (12.37) and (12.38) can be combined to give a single complex equation for s:

$$\ddot{s} = -\frac{g}{l}\,s - i\,2\,\Omega\,\sin\lambda\,\dot{s}. \tag{12.40}$$

Let us look for a sinusoidally oscillating solution of the form

$$s = s_0\,e^{-i\,\omega\,t}. \tag{12.41}$$

Here, ω is the (real) angular frequency of oscillation and s_0 is an arbitrary complex constant. Equations (12.40) and (12.41) yield the following quadratic equation for ω:

$$\omega^2 - 2\,\Omega\,\sin\lambda\,\omega - \frac{g}{l} = 0. \tag{12.42}$$

The solutions are approximately

$$\omega_\pm \simeq \Omega\,\sin\lambda \pm \sqrt{\frac{g}{l}}, \tag{12.43}$$

where we have neglected terms involving Ω^2. Hence, the general solution of (12.41) takes the form

$$s = s_+\,e^{-i\,\omega_+\,t} + s_-\,e^{-i\,\omega_-\,t}, \tag{12.44}$$

where s_+ and s_- are two arbitrary complex constants.

Making the specific choice $s_+ = s_- = a/2$, where a is real, the previous solution reduces to

$$s = a\,e^{-i\,\Omega\,\sin\lambda\,t}\,\cos\left(\sqrt{\frac{g}{l}}\,t\right), \tag{12.45}$$

where we have employed Euler's theorem, $e^{i\theta} = \cos\theta + i\sin\theta$. It is clear from Equation (12.39) that x' and y' are the real and imaginary parts of s, respectively. Thus, it follows from the previous equation that

$$x' = a\,\cos(\Omega\,\sin\lambda\,t)\,\cos\left(\sqrt{\frac{g}{l}}\,t\right),\tag{12.46}$$

$$y' = -a\,\sin(\Omega\,\sin\lambda\,t)\,\cos\left(\sqrt{\frac{g}{l}}\,t\right),\tag{12.47}$$

where we have again employed Euler's theorem. These equations describe sinusoidal oscillations, in a plane whose normal is parallel to the Earth's surface, at the standard pendulum angular frequency $(g/l)^{1/2}$. (See Section 11.4.) The Coriolis force, however, causes the plane of oscillation to slowly precess at the angular frequency $\Omega\sin\lambda$. The period of the precession is

$$T = \frac{2\pi}{\Omega\,\sin\lambda} \simeq \frac{24}{\sin\lambda}\ \mathrm{hrs}.\tag{12.48}$$

For example, according to the previous equations, the pendulum oscillates in the x'-direction (i.e., north/south) at $t \simeq 0$, in the y'-direction (i.e., east/west) at $t \simeq T/4$, in the x'-direction again at $t \simeq T/2$, et cetera. The precession is clockwise (looking from above) in the northern hemisphere and counter-clockwise in the southern hemisphere.

The precession of the plane of oscillation of a pendulum, due to the Coriolis force, is used in many museums and observatories to demonstrate that the Earth is rotating. This method of making the Earth's rotation manifest was first devised by Léon Foucault in 1851.

12.6 EXERCISES

12.1 A pebble is dropped down an elevator shaft in the Empire State Building ($h = 1250$ ft, latitude $= 41°$ N). Neglect air resistance.

  (a) What is the magnitude of the pebble's horizontal deflection due to the Coriolis force at the bottom of the shaft? [Ans: 12.3 cm.]

  (b) What is the direction of the deflection? [Ans: Eastward.]

  [Modified from Fowles & Cassiday 2005.]

12.2 If a bullet is fired due east, at an elevation angle α, from a point on the Earth whose latitude is $+\lambda$. Let Ω be the Earth's angular velocity, v_0 the bullet's initial speed, and g is the acceleration due to gravity. Neglect air resistance.

  (a) Show that the bullet will strike the Earth with a lateral deflection $4\,\Omega\,v_0^3\,\sin\lambda\,\sin^2\alpha\,\cos\alpha/g^2$.

  (b) Is the deflection northward or southward? [Ans: Southward.]

  [Modified from Thornton & Marion 2004.]

12.3 A particle is thrown vertically with initial speed v_0, reaches a maximum height, and falls back to the ground. Neglect air resistance. Show that the horizontal Coriolis deflection of the particle when it returns to the ground is opposite in direction, and four times greater in magnitude, than the Coriolis deflection when it is dropped at rest from the same maximum height. [From Goldstein et al. 2001.]

12.4 A ball of mass m rolls without friction over a horizontal plane located on the surface of the Earth. Show that in the northern hemisphere it rolls in a clockwise sense (seen from above) around a circle of radius

$$r = \frac{v}{2\,\Omega\,\sin\lambda},$$

where v is the speed of the ball, Ω the Earth's angular velocity, and λ the terrestrial latitude. [From Fitzpatrick 2012.]

12.5 Demonstrate that the Coriolis force causes conical pendulums to rotate clockwise and counter-clockwise at slightly different angular frequencies. What is the frequency difference as a function of terrestrial latitude? [Ans: $2\,\Omega\,\sin\lambda$.]

12.6 A satellite is in a circular orbit of radius a about the Earth. Let us define a set of co-moving Cartesian coordinates, centered on the satellite, such that the x-axis always points toward the center of the Earth, the y-axis in the direction of the satellite's orbital motion, and the z-axis in the direction of the satellite's orbital angular velocity, $\boldsymbol{\omega}$.

(a) Demonstrate that the equation of motion of a small mass in orbit about the satellite are

$$\ddot{x} = 3\,\omega^2\,x + 2\,\omega\,\dot{y},$$
$$\ddot{y} = -2\,\omega\,\dot{x},$$

assuming that $|x|/a \ll 1$ and $|y|/a \ll 1$. You may neglect the gravitational attraction between the satellite and the mass.

(b) Show that the mass executes a retrograde (i.e., in the opposite sense to the satellite's orbital rotation) elliptical orbit about the satellite whose period matches that of the satellite's orbit, and whose major and minor axes are in the ratio $2:1$, and are aligned along the y- and x-axes, respectively.

[From Fitzpatrick 2012.]

CHAPTER 13

Newtonian Gravity

13.1 INTRODUCTION

In this chapter, we shall examine Newton's theory of **gravity**.

13.2 UNIVERSAL GRAVITY

The universal law of gravity was first correctly described in Newton's *Principia*. According to Newton, any two point objects exert a gravitational force of attraction on one another; this force is directed along the straight-line joining the objects, is directly proportional to the product of their masses, and is inversely proportional to the square of the distance between them.

Consider two point objects of masses m_1 and m_2 that are separated by a distance r. The magnitude of the gravitational force of attraction acting between the objects is

$$f = G\,\frac{m_1\,m_2}{r^2}, \tag{13.1}$$

where the constant of proportionality, G, is called the *universal gravitational constant*, and takes the value

$$G = 6.674 \times 10^{-11}\,\mathrm{m^3\,kg^{-1}\,s^{-2}} \tag{13.2}$$

(Yoder 1995). The universal gravitational constant is numerically very small, which implies that gravity is an intrinsically weak force. In fact, gravity only usually becomes significant when at least one of the masses involved is of astronomical dimensions (i.e., is a planet or a star).

Incidentally, there is something rather curious about Equation (13.1). According to this equation, the gravitational force acting on a given object is directly proportional to that object's inertial mass. Why, though, should inertia be related to the force of gravity? After all, inertia simply measures the reluctance of a given body to deviate from its preferred state of uniform motion in a straight-line, in response to some external force. The preceding question perplexed physicists for many years, but was eventually answered when Albert Einstein published his *general theory of relativity* in 1915. According to Einstein, inertial mass acts as a sort of gravitational charge because it is impossible to distinguish an acceleration produced by a gravitational field from an apparent acceleration generated by observing motion in a non-inertial reference frame. The assumption that these two types of acceleration are indistinguishable leads directly to all of the strange predictions of general relativity; for instance, that clocks in different gravitational potentials run at different rates, that mass bends space, and so on.

DOI: 10.1201/9781003198642-13

13.2.1 Surface Gravity

Let us use Newton's universal law of gravity to account for the Earth's surface gravity. Consider an object of mass m that is located very close to the Earth's surface. Now, the Earth's mean radius is $R_{\text{earth}} = 6.371 \times 10^6$ m, whereas its mass is $M_{\text{earth}} = 5.972 \times 10^{24}$ kg (Yoder 1995). Newton proved, after considerable effort, that the gravitational force acting outside a spherically-symmetric mass distribution is the same as that due to an equivalent point mass located at the geometric center of that distribution. (See Section 15.4.) Hence, the gravitational force exerted by the (almost) spherical Earth on the object in question is of magnitude

$$f = \frac{G m M_{\text{earth}}}{R_{\text{earth}}^2}, \tag{13.3}$$

and is directed toward the center of the Earth. It follows that the object is subject to a net downward acceleration of magnitude

$$g_{\text{earth}} = \frac{f}{m} = \frac{G M_{\text{earth}}}{R_{\text{earth}}^2} = \frac{(6.674 \times 10^{-11}) \times (5.972 \times 10^{24})}{(6.371 \times 10^6)^2} = 9.82 \, \text{m/s}^2 \tag{13.4}$$

that is independent of its mass. This estimate for the acceleration due to gravity differs very slightly from the conventional value for the mean acceleration ($9.81 \, \text{m/s}^2$) because of the centrifugal effect of the Earth's diurnal rotation (see Section 12.3), as well as the fact that the Earth is not quite spherical. (See Section 15.6.)

Because Newton's law of gravitation is universal, we immediately conclude that any spherical body of mass M and radius R possesses a surface acceleration due to gravity, g, given by the following formula:

$$\frac{g}{g_{\text{earth}}} = \frac{M/M_{\text{earth}}}{(R/R_{\text{earth}})^2}. \tag{13.5}$$

Table 13.1 shows the surface acceleration due to gravity of various bodies in the solar system, estimated using the previous expression. It can be seen that the surface gravity of the Moon is only about one fifth of that of the Earth. No wonder Apollo astronauts were able to jump so far on the Moon's surface! Prospective Mars colonists should note that they will only weigh about a third of their terrestrial weight on Mars.

13.3 GRAVITATIONAL POTENTIAL ENERGY

We saw earlier, in Section 5.5, that since gravity is a conservative force, it must have an associated potential energy. Let us obtain a general formula for this energy. Consider a point object of mass m that is located at a radial distance r from another point object of mass M. The gravitational force acting on the first mass is of magnitude $f = G m M/r^2$, and is directed toward the second mass. Imagine that the first mass moves radially away from the second mass until it reaches infinity. What is the change in the potential energy of the first mass associated with this shift in position? According to Equation (5.34),

$$U(\infty) - U(r) = -\int_r^\infty [-f(r)] \, dr. \tag{13.6}$$

There is a minus sign in front of f because this force is oppositely directed to the motion. The previous expression can be integrated to give

$$U(r) = -\frac{G m M}{r}. \tag{13.7}$$

Table 13.1 The Mass, M, Radius, R, Surface Gravity, g, and Escape Velocity, u, of Various Bodies in the Solar System. All Quantities are Expressed as Fractions of the Corresponding terrestrial Quantity (All data are derived from Yoder 1995.)

Body	M/M_{earth}	R/R_{earth}	g/g_{earth}	u/u_{earth}
Sun	3.33×10^5	109.0	28.1	55.3
Moon	0.0123	0.273	0.17	0.21
Mercury	0.0553	0.383	0.38	0.38
Venus	0.816	0.949	0.91	0.93
Earth	1.000	1.000	1.000	1.000
Mars	0.108	0.533	0.38	0.45
Jupiter	318.3	11.21	2.5	5.3
Saturn	95.14	9.45	1.07	3.2
Uranus	14.54	3.98	0.92	1.9
Neptune	17.15	3.86	1.15	2.1

Here, we have adopted the convenient normalization that the potential energy at infinity is zero. According to the previous formula, the gravitational potential energy of a point mass m located a distance r from a point mass M is simply $-G\,m\,M/r$. Incidentally, the fact that the gravitational force acting outside a spherically-symmetric mass distribution is the same as that due to an equivalent point mass located at the geometric center of the distribution implies that if a mass m is located a distance $r > R$ from a spherically-symmetric mass distribution of mass M, and radius R, then the mass's gravitational potential energy is also $-G\,m\,M/r$.

Consider an object of mass m moving relatively close to the Earth's surface. The potential energy of such an object can be written

$$U = -\frac{G\,m\,M_{\text{earth}}}{R_{\text{earth}} + z}. \tag{13.8}$$

where z is the vertical height of the object above the Earth's surface. In the limit that $z \ll R_{\text{earth}}$, the previous expression can be expanded to give

$$U \simeq -\frac{G\,m\,M_{\text{earth}}}{R_{\text{earth}}} + \frac{G\,m\,M_{\text{earth}}}{R_{\text{earth}}^2}\,z, \tag{13.9}$$

Because potential energy is undetermined to an arbitrary additive constant (see Section 5.5), we could just as well write

$$U \simeq m\,g_{\text{earth}}\,z, \tag{13.10}$$

where $g_{\text{earth}} = G\,M_{\text{earth}}/R_{\text{earth}}^2$ is the acceleration due to gravity at the Earth's surface. [See Equation (13.4).] Of course, the previous expression is equivalent to the expression (5.4) derived earlier in this book.

13.3.1 Escape Velocity

Consider a relatively small object of mass m and speed v, moving in the gravitational field of a spherically-symmetric mass distribution of mass M. We expect the object's total energy,

$$E = K + U, \tag{13.11}$$

to be a constant of the motion. Here, the kinetic energy is written $K = (1/2)\,m\,v^2$, whereas the potential energy takes the form $U = -G\,m\,M/r$. Of course, r is the distance between the object and the center of the mass distribution. Suppose that the mass distribution is a sphere of outer radius R. Suppose, further, that the small object is launched from the surface of the sphere with some speed, u, which is such that it only just escapes the sphere's gravitational influence. After the object has escaped, it is a long way from the sphere, and, hence, $U = 0$. Moreover, if the object only just managed to escape then we also expect $K = 0$, because the object will have expended all of its initial kinetic energy escaping from the sphere's gravitational well. We conclude that the launched object, after it has escaped from the gravitational influence of the sphere, possesses zero net energy; that is, $E = K + U = 0$. Because E is a constant of the motion, it follows that at the launch point

$$E = \frac{1}{2}\,m\,u^2 - \frac{G\,m\,M}{R} = 0. \tag{13.12}$$

This expression can be rearranged to give

$$u = \sqrt{\frac{2\,G\,M}{R}}. \tag{13.13}$$

The quantity u is known as the *escape velocity* (actually, it is a speed). An object launched from the surface of the sphere with a speed in excess of this value can eventually escape from the sphere's gravitational influence. Otherwise, the object must remain in orbit around the sphere, and may eventually strike its surface. Note that the escape velocity is independent of the object's mass and launch direction (assuming that it is not into the sphere).

The escape velocity for the Earth is

$$u_{\text{earth}} = \sqrt{\frac{2\,G\,M_{\text{earth}}}{R_{\text{earth}}}} = \sqrt{\frac{2 \times (6.674 \times 10^{-11}) \times (5.972 \times 10^{24})}{6.371 \times 10^6}} = 11.2\,\text{km/s}. \tag{13.14}$$

The escape velocity for a general spherical body of mass M and radius R can be written

$$\frac{u}{u_{\text{earth}}} = \sqrt{\frac{M/M_{\text{earth}}}{R/R_{\text{earth}}}}. \tag{13.15}$$

Table 13.1 shows the escape velocities of various bodies in the solar system, estimated using the previous expression. Generally speaking, bodies larger than the Earth have higher escape velocities than the Earth, and vice versa.

13.4 CIRCULAR ORBITS

Consider an object executing a circular orbit of radius r around the Earth. Let ω be the object's orbital angular velocity. The object experiences an acceleration toward the Earth's center of magnitude $\omega^2 r$. (See Section 7.3.) Of course, this acceleration is provided by the gravitational attraction between the object and the Earth, which yields an acceleration of magnitude $G\,M_{\text{earth}}/r^2$. It follows that

$$\omega^2 r = \frac{G\,M_{\text{earth}}}{r^2}. \tag{13.16}$$

13.4.1 Lunar Orbital Period

Let us use the previous equation to reproduce a calculation, first performed by Isaac Newton, that demonstrates that the same force of gravity that causes objects close to the Earth's surface to accelerate downward at $9.81\,\mathrm{m/s}^2$ is also responsible for maintaining the Moon in its orbit around the Earth. Let r_{moon} be the mean radius of the Moon's approximately circular orbit around the Earth, and let R_{earth} be the Earth's mean radius. It is possible to accurately measure the ratio $r_{\mathrm{moon}}/R_{\mathrm{earth}}$ by observing the parallax of the Moon (i.e., the apparent change in position of the Moon in the sky seen by an observer on the surface of the Earth whose position changes due to the Earth's diurnal rotation). In fact, as was well known in Newton's day, $r_{\mathrm{moon}}/R_{\mathrm{earth}} = 60.3$ (Yoder 1995). Let $T_{\mathrm{moon}} = 2\pi/\omega$ be the Moon's (sidereal) orbital period around the Earth. Equation (13.16) can be rearranged to give

$$T_{\mathrm{moon}} = 2\pi \left(\frac{R_{\mathrm{earth}}}{g_{\mathrm{earth}}}\right)^{1/2} \left(\frac{r_{\mathrm{moon}}}{R_{\mathrm{earth}}}\right)^{3/2}, \tag{13.17}$$

where $g_{\mathrm{earth}} = G\,M_{\mathrm{earth}}/R_{\mathrm{earth}}^2$ is the Earth's surface gravity. Hence,

$$T_{\mathrm{moon}} = 2\pi \left(\frac{6.371 \times 10^6}{9.81}\right)^{1/2} \times (60.3)^{3/2} = 27.4\,\mathrm{days}. \tag{13.18}$$

Note, incidentally, that the values of both g_{earth} and R_{earth} were well known in Newton's day. Our estimate for the sidereal (i.e., relative to the stars, rather than the Sun) orbital period of the Moon is very close to the observed value (27.3 days). Hence, we have indeed demonstrated that the force of gravity can account for both the acceleration of objects close to the Earth's surface and the orbit of the Moon.

13.4.2 Geostationary Satellites

Consider an artificial satellite in orbit around the Earth. Suppose that the satellite's orbit lies in the Earth's equatorial plane. Moreover, suppose that the satellite's orbital angular velocity exactly matches the Earth's angular velocity of diurnal rotation. In this case, the satellite will appear to hover in the same place in the sky to a stationary observer on the Earth's surface. A satellite with this singular property is known as a *geostationary satellite*.

Virtually all of the satellites used to monitor the Earth's weather patterns are geostationary in nature. Communications satellites also tend to be geostationary. Of course, the satellites that beam satellite-TV into homes across the world must be geostationary; otherwise, you would need to install an expensive tracking antenna on top of your house in order to pick up the transmissions. Incidentally, the person who first envisaged rapid global telecommunication via a network of geostationary satellites was the science fiction writer Arthur C. Clarke in 1945.

Let us calculate the orbital radius of a geostationary satellite. The angular velocity of the Earth's diurnal rotation is

$$\Omega = 7.2921 \times 10^{-5}\,\mathrm{rad/s}. \tag{13.19}$$

[See Equation (12.10).] It follows from Equation (13.16) that

$$r_{\mathrm{geo}} = \left(\frac{G\,M_{\mathrm{earth}}}{\Omega^2}\right)^{1/3} = \left[\frac{(6.674 \times 10^{-11}) \times (5.972 \times 10^{24})}{(7.2921 \times 10^{-5})^2}\right]^{1/3}$$

$$= 4.216 \times 10^7\,\mathrm{m} = 6.62\,R_{\mathrm{earth}}. \tag{13.20}$$

Thus, a geostationary satellite must be placed in a circular orbit whose radius is exactly 6.62 times the Earth's mean radius.

13.5 EXERCISES

13.1 A particle is projected vertically upward from the Earth's surface with a velocity that would, were gravity uniform, carry it to a height h. Show that if the variation of gravity with height is allowed for, but the resistance of air is neglected, then the height reached will be greater by $h^2/(R-h)$, where R is the Earth's radius. [From Lamb 1942.]

13.2 A particle is projected vertically upward from the Earth's surface with a velocity just sufficient for it to reach infinity (neglecting air resistance). Prove that the time needed to reach a height h is

$$\frac{1}{3}\left(\frac{2R}{g}\right)^{1/2}\left[\left(1+\frac{h}{R}\right)^{3/2}-1\right],$$

where R is the Earth's radius and g its surface gravitational acceleration. [From Lamb 1942.]

13.3 Assuming that the Earth is a sphere of radius R, and neglecting air resistance, show that a particle that starts from rest a distance R from the Earth's surface will reach the surface with speed $\sqrt{Rg}$ after a time $(1+\pi/2)\sqrt{R/g}$, where g is the surface gravitational acceleration.

13.4 A rocket is located a distance 3.5 times the radius of the Earth above the Earth's surface. What is the rocket's free-fall acceleration? [Ans: $0.484\,\mathrm{m/s^2}$.]

13.5 Callisto is Jupiter's eighth moon; its mass and radius are $M=1.08\times10^{23}\,\mathrm{kg}$ and $R=2403\,\mathrm{km}$, respectively. What is the gravitational acceleration on the surface of Callisto? [Ans: $1.25\,\mathrm{m/s^2}$.]

13.6 What is the minimum energy required to launch a probe of mass $m=120\,\mathrm{kg}$ into outer space from the surface of the Earth? [Ans: $7.495\times10^9\,\mathrm{J}$.]

13.7 A satellite moves in a circular orbit around the Earth with speed $v=6000\,\mathrm{m/s}$.

(a) Determine the satellite's altitude above the Earth's surface. [Ans: $4.69\times10^6\,\mathrm{m}$.]

(b) Determine the period of the satellite's orbit. [Ans: 3.22 hours.]

13.8 A planet is in circular orbit around a star. The period and radius of the orbit are $T=4.3\times10^7\,\mathrm{s}$ and $r=2.34\times10^{11}\,\mathrm{m}$, respectively. Calculate the mass of the star. [Ans: $4.01\times10^{30}\,\mathrm{kg}$.]

13.9 Given the Sun's mean apparent radius seen from the Earth (16′), the Earth's mean apparent radius seen from the Moon (57′), and the mean number of lunar revolutions in a year (13.4), show that the ratio of the Sun's mean density to that of the Earth is 0.252. [From Lamb 1942.]

13.10 Prove that the orbital period of a satellite close to the surface of a spherical planet depends on the mean density of the planet, but not on its size. Show that if the mean density is that of water then the period is 3 h, 18 m. [From Lamb 1942.]

13.11 Jupiter's satellite Ganymede has an orbital period of 7 d, 3 h, 43 m, and a mean orbital radius that is 15.3 times the mean radius of the planet. The Moon has an orbital period of 27 d, 7 h, 43 m, and a mean orbital radius that is 60.3 times the Earth's mean radius. Show that the ratio of Jupiter's mean density to that of the Earth is 0.238. [From Lamb 1942.]

13.12 Show that the velocity acquired by a particle falling from a great distance into the Sun is $\sqrt{2V^2/a}$, where V is the Earth's orbital velocity, and a is the apparent angular radius of the Sun as seen from the Earth.

13.13 A mass falls vertically from rest toward the surface of the Earth. The initial (at $t = 0$) distance of the mass from the center of the Earth is r_0, which is assumed to be much larger than the Earth's radius. Neglect air resistance.

    (a) If r is the distance of the mass from the center of the Earth at time t, show that

$$r_0^{3/2} \left(\sqrt{\frac{r}{r_0} \left(1 - \frac{r}{r_0} \right)} + \sin^{-1} \sqrt{1 - \frac{r}{r_0}} \right) = \sqrt{GM}\, t,$$

    where M is the mass of the Earth.

    (b) Hence, deduce that the mass required approximately 9/11 of the total time of fall to traverse the first half of the distance through which it falls.

13.14 Prove that two equal gravitating spheres, of radius a, and density equal to the Earth's mean density, starting from rest a great distance apart, and subject only to their mutual attraction, will collide with a velocity $\sqrt{g\,a^2/(2\,R)}$, where R is the Earth's radius, and g is its surface gravitational acceleration. [From Lamb 1942.]

13.15 Two gravitating masses m_1 and m_2 are separated by a distance r_0 and released from rest. Show that when the separation is r (where $r < r_0$) the respective speeds are

$$v_1 = m_2 \sqrt{\frac{2\,G}{M} \left(\frac{1}{r} - \frac{1}{r_0} \right)},$$

$$v_2 = m_1 \sqrt{\frac{2\,G}{M} \left(\frac{1}{r} - \frac{1}{r_0} \right)},$$

where $M = m_1 + m_2$.

CHAPTER 14

Orbital Motion

14.1 INTRODUCTION

One of the most famous applications of Newtonian dynamics is the analysis of **planetary motion** around the Sun. Let us now investigate this problem.

14.2 KEPLER'S LAWS

As is well known, Johannes Kepler was the first astronomer to correctly describe planetary motion in the solar system (in works published between 1609 and 1619). The motion of the planets is summed up in three simple laws:

1. The planetary orbits are all ellipses that are confocal with the Sun (i.e., the Sun lies at one of the focii of each ellipse).

2. The radius vectors connecting each planet to the Sun sweep out equal areas in equal time intervals.

3. The squares of the orbital periods of the planets are proportional to the cubes of their orbital major radii. (See Table 14.1.)

Let us now see if we can derive Kepler's laws from Newton's laws of motion.

14.3 PLANETARY EQUATIONS OF MOTION

Consider the motion of a planet around the Sun. Suppose that the Sun, which is of mass M, is located at the origin of our coordinate system. Suppose that the planet, which is of mass m, is located at position vector $\mathbf{r}$. The gravitational force exerted on the planet by the Sun is written

$$\mathbf{f} = -\frac{G M m}{r^3} \mathbf{r}. \tag{14.1}$$

[This is just a vector version of Equation (13.1).] An equal and opposite force to (14.1) acts on the Sun. However, we shall assume that the Sun is so much more massive than the planet in question that this force does not cause the Sun's position to shift appreciably. Hence, the Sun will always remain at the origin of our coordinate system. Likewise, we shall neglect the gravitational forces exerted on our planet by the other planets in the solar system, compared to the much larger gravitational force exerted by the Sun. Newton's second law of motion yields

$$m \ddot{\mathbf{r}} = \mathbf{f}, \tag{14.2}$$

DOI: 10.1201/9781003198642-14

which reduces to

$$\ddot{\mathbf{r}} = -\frac{GM}{r^3}\mathbf{r}. \tag{14.3}$$

Note that the planetary equation of motion is independent of the planetary mass.

Gravity is a conservative force. Hence, the gravitational force (14.1) can be written

$$\mathbf{f} = -\nabla U, \tag{14.4}$$

where the potential energy, $U(\mathbf{r})$, of our planet in the Sun's gravitational field takes the form

$$U(\mathbf{r}) = -\frac{GMm}{r}. \tag{14.5}$$

(See Section 13.3.) It follows that the total energy of our planet is a conserved quantity. (See Section 5.5.) In other words,

$$\mathcal{E} = \frac{v^2}{2} - \frac{GM}{r} \tag{14.6}$$

is constant in time. Here, $\mathcal{E}$ is actually the planet's total energy per unit mass, and $\mathbf{v} = d\mathbf{r}/dt$.

Gravity is also a central force. Hence, the angular momentum of our planet (about the Sun) is a conserved quantity. (See Section 9.4.) In other words,

$$\mathbf{h} = \mathbf{r} \times \mathbf{v}, \tag{14.7}$$

which is actually the planet's angular momentum per unit mass, is constant in time. Taking the scalar product of the previous equation with $\mathbf{r}$, we obtain

$$\mathbf{h} \cdot \mathbf{r} = 0. \tag{14.8}$$

This is the equation of a plane that passes through the origin, and whose normal is parallel to $\mathbf{h}$. Because $\mathbf{h}$ is a constant vector, it always points in the same direction. We, therefore, conclude that the motion of our planet is two-dimensional in nature; that is, it is confined to some fixed plane that passes through the origin. Without loss of generality, we can let this plane coincide with the x-y plane.

It is convenient to specify the instantaneous position of our planet in the x-y plane in terms of the polar coordinates r and θ, as illustrated in Figure 7.6. Note that $x = r\cos\theta$ and $y = r\sin\theta$. Recall, from Section 7.7, that the radial and tangential components of the planet's velocity can be written

$$v_r = \dot{r}, \tag{14.9}$$

$$v_\theta = r\dot{\theta}, \tag{14.10}$$

respectively. Likewise, the radial and tangential components of the planet's acceleration take the respective forms

$$a_r = \ddot{r} - r\dot{\theta}^2, \tag{14.11}$$

$$a_\theta = r\ddot{\theta} + 2\dot{r}\dot{\theta}. \tag{14.12}$$

Finally, the radial and tangential components of the gravitational force (14.1) exerted on the planet by the Sun are

$$f_r = -\frac{GMm}{r^2}, \tag{14.13}$$

$$f_\theta = 0, \tag{14.14}$$

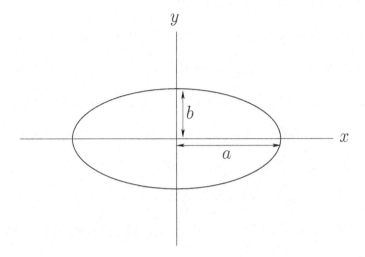

Figure 14.1 An ellipse. (Reproduced from Fitzpatrick 2012. Courtesy of Cambridge University Press.)

respectively.

When expressed in terms of polar coordinates, Newton's second law of motion for the planet yields

$$m\, a_r = f_r, \tag{14.15}$$

$$m\, a_\theta = f_\theta. \tag{14.16}$$

Hence, we obtain the planet's radial equation of motion,

$$\ddot{r} - r\, \dot{\theta}^{\,2} = -\frac{G\,M}{r^{\,2}}, \tag{14.17}$$

as well as its tangential equation of motion,

$$r\, \ddot{\theta} + 2\, \dot{r}\, \dot{\theta} = 0. \tag{14.18}$$

The previous two equations are the polar components of the vector equation (14.3).

14.4 CONIC SECTIONS

The ellipse, the parabola, and the hyperbola are collectively known as *conic sections*, because these three types of curve can be obtained by taking various different plane sections of a right cone. It turns out that the possible solutions of Equations (14.17) and (14.18) are all conic sections. It is, therefore, appropriate for us to briefly review these curves.

An *ellipse*, centered on the origin, of major radius a and minor radius b, which are aligned along the x- and y-axes, respectively (see Figure 14.1), satisfies the following well-known equation:

$$\frac{x^{\,2}}{a^{\,2}} + \frac{y^{\,2}}{b^{\,2}} = 1. \tag{14.19}$$

Likewise, a *parabola* that is aligned along the $+x$-axis, and passes through the origin (see Figure 14.2), satisfies:

$$y^{\,2} - b\,x = 0, \tag{14.20}$$

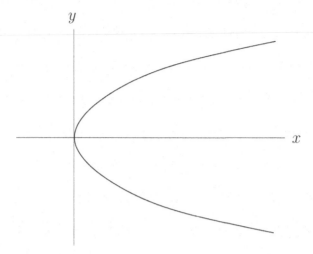

Figure 14.2 A parabola. (Reproduced from Fitzpatrick 2012. Courtesy of Cambridge University Press.)

where $b > 0$.

Finally, a *hyperbola* that is aligned along the $+x$-axis, and whose asymptotes intersect at the origin (see Figure 14.3), satisfies:

$$\frac{x^2}{a^2} - \frac{y^2}{b^2} = 1. \tag{14.21}$$

Here, a is the distance of closest approach to the origin. The asymptotes subtend an angle $\phi = \tan^{-1}(b/a)$ with the x-axis.

It is not clear, at this stage, what the ellipse, the parabola, and the hyperbola have in common (other than being conic sections). It turns out that what these three curves have in common is that they can all be represented as the locus of a movable point whose distance from a fixed point is in a constant ratio to its perpendicular distance to some fixed straight-line. Let the fixed point (which is termed the *focus* of the ellipse/parabola/hyperbola) lie at the origin, and let the fixed straight-line correspond to $x = -d$ (with $d > 0$). Thus, the distance of a general point (x, y) (which lies to the right of the line $x = -d$) from the origin is $r_1 = (x^2 + y^2)^{1/2}$, whereas the perpendicular distance of the point from the line $x = -d$ is $r_2 = x + d$. See Figure 14.4. In polar coordinates, $r_1 = r$ and $r_2 = r \cos \theta + d$. Hence, the locus of a point for which r_1 and r_2 are in a fixed ratio satisfies the following equation:

$$\frac{r_1}{r_2} = \frac{\sqrt{x^2 + y^2}}{x + d} = \frac{r}{r \cos \theta + d} = e, \tag{14.22}$$

where $e \geq 0$ is a constant. When expressed in terms of polar coordinates, the previous equation can be rearranged to give

$$r = \frac{r_c}{1 - e \cos \theta}, \tag{14.23}$$

where $r_c = e\, d$.

When written in terms of Cartesian coordinates, Equation (14.22) can be rearranged to give

$$\frac{(x - x_c)^2}{a^2} + \frac{y^2}{b^2} = 1, \tag{14.24}$$

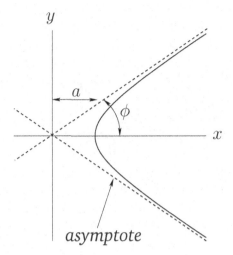

Figure 14.3 A hyperbola. (Reproduced from Fitzpatrick 2012. Courtesy of Cambridge University Press.)

for $e < 1$. Here,

$$a = \frac{r_c}{1 - e^2}, \tag{14.25}$$

$$b = \frac{r_c}{\sqrt{1 - e^2}} = \sqrt{1 - e^2}\, a, \tag{14.26}$$

$$x_c = \frac{e\, r_c}{1 - e^2} = e\, a. \tag{14.27}$$

Equation (14.24) can be recognized as the equation of an ellipse whose center lies at $(x_c,\, 0)$, and whose major and minor radii, a and b, are aligned along the x- and y-axes, respectively [cf., Equation (14.19)].

When again written in terms of Cartesian coordinates, Equation (14.22) can be rearranged to give

$$y^2 - 2\, r_c\, (x - x_c) = 0, \tag{14.28}$$

for $e = 1$. Here, $x_c = -r_c/2$. This is the equation of a parabola that passes through the point $(x_c,\, 0)$, and that is aligned along the $+x$-direction [cf., Equation (14.20)].

Finally, when written in terms of Cartesian coordinates, Equation (14.22) can be rearranged to give

$$\frac{(x - x_c)^2}{a^2} - \frac{y^2}{b^2} = 1, \tag{14.29}$$

for $e > 1$. Here,

$$a = \frac{r_c}{e^2 - 1}, \tag{14.30}$$

$$b = \frac{r_c}{\sqrt{e^2 - 1}} = \sqrt{e^2 - 1}\, a, \tag{14.31}$$

$$x_c = -\frac{e\, r_c}{e^2 - 1} = -e\, a. \tag{14.32}$$

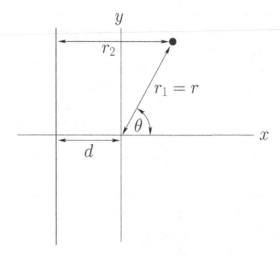

Figure 14.4 Conic sections in polar coordinates. (Reproduced from Fitzpatrick 2012. Courtesy of Cambridge University Press.)

Equation (14.29) can be recognized as the equation of a hyperbola whose asymptotes intersect at $(x_c,\,0)$, and that is aligned along the $+x$-direction. The asymptotes subtend an angle

$$\phi = \tan^{-1}\left(\frac{b}{a}\right) = \tan^{-1}\left(\sqrt{e^2 - 1}\right) \tag{14.33}$$

with the x-axis [cf., Equation (14.21)].

In conclusion, Equation (14.23) is the polar equation of a general conic section that is confocal with the origin. For $e < 1$, the conic section is an ellipse. For $e = 1$, the conic section is a parabola. Finally, for $e > 1$, the conic section is a hyperbola.

14.5 KEPLER'S SECOND LAW

Multiplying our planet's tangential equation of motion, (14.18), by r, we obtain

$$r^2\,\ddot{\theta} + 2\,r\,\dot{r}\,\dot{\theta} = 0. \tag{14.34}$$

However, the previous equation can also be written

$$\frac{d(r^2\,\dot{\theta})}{dt} = 0, \tag{14.35}$$

which implies that

$$h = r^2\,\dot{\theta} \tag{14.36}$$

is constant in time. It is easily demonstrated that h is the magnitude of the vector $\mathbf{h}$ defined in Equation (14.7). Thus, the fact that h is constant in time is equivalent to the statement that the angular momentum of our planet is a constant of its motion. As we have already mentioned, this is the case because gravity is a central force.

Suppose that the radius vector connecting our planet to the origin (i.e., the Sun) sweeps out an angle $\delta\theta$ between times t and $t + \delta t$. See Figure 14.5. The approximately triangular

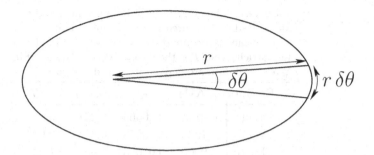

Figure 14.5 Kepler's second law. (Reproduced from Fitzpatrick 2012. Courtesy of Cambridge University Press.)

region swept out by the radius vector has the area

$$\delta A \simeq \frac{1}{2} r^2 \delta\theta, \tag{14.37}$$

because the area of a triangle is half its base $(r\,\delta\theta)$ times its height (r). Hence, the rate at which the radius vector sweeps out area is

$$\frac{dA}{dt} = \lim_{\delta t \to 0} \frac{r^2\,\delta\theta}{2\,\delta t} = \frac{r^2}{2}\frac{d\theta}{dt} = \frac{h}{2}. \tag{14.38}$$

Thus, the radius vector sweeps out area at a constant rate (because h is constant in time). This is Kepler's second law. We conclude that Kepler's second law of planetary motion is a direct consequence of angular momentum conservation.

14.6 KEPLER'S FIRST LAW

Our planet's radial equation of motion, (14.17), can be combined with Equation (14.36) to give

$$\ddot{r} - \frac{h^2}{r^3} = -\frac{G\,M}{r^2}. \tag{14.39}$$

Suppose that $r = u^{-1}$, where $u = u(\theta)$. It follows that

$$\dot{r} = -\frac{\dot{u}}{u^2} = -r^2\frac{du}{d\theta}\frac{d\theta}{dt} = -h\frac{du}{d\theta}. \tag{14.40}$$

Likewise,

$$\ddot{r} = -h\frac{d^2u}{d\theta^2}\,\dot{\theta} = -u^2\,h^2\frac{d^2u}{d\theta^2}. \tag{14.41}$$

Hence, Equation (14.39) can be written in the linear form

$$\frac{d^2u}{d\theta^2} + u = \frac{G\,M}{h^2}. \tag{14.42}$$

The general solution to the previous equation is

$$u(\theta) = \frac{G\,M}{h^2}\left[1 - e\,\cos(\theta - \theta_0)\right], \tag{14.43}$$

Table 14.1 Planetary Orbital Parameters. a is the
Major Radius Measured in Astronomical units (1 AU
is the mean Earth-Sun distance), e is the Orbital
Eccentricity, and T is the Orbital Period Measured in
Years. (All data are derived from Yoder 1995.)

Planet	a(AU)	e	T(yr)	a^3/T^2
Mercury	0.3871	0.20564	0.241	0.999
Venus	0.7233	0.00676	0.615	0.994
Earth	1.0000	0.01673	1.000	1.000
Mars	1.5237	0.09337	1.881	1.000
Jupiter	5.2025	0.04854	11.862	1.001
Saturn	9.5415	0.05551	29.457	1.001
Uranus	19.188	0.04686	84.021	1.001
Neptune	30.070	0.00895	164.8	1.001

where e and θ_0 are arbitrary constants. Without loss of generality, we can set $\theta_0 = 0$ by rotating our coordinate system about the z-axis. Thus, we obtain

$$r(\theta) = \frac{r_c}{1 - e \cos\theta},$$

(14.44)

where

$$r_c = \frac{h^2}{G M}.$$

(14.45)

We immediately recognize Equation (14.44) as the equation of a conic section that is confocal with the origin (i.e., with the Sun). Specifically, for $e < 1$, Equation (14.44) is the equation of an ellipse that is confocal with the Sun. Thus, the orbit of our planet around the Sun in a confocal ellipse. This is Kepler's first law of planetary motion. Of course, a planet cannot have a parabolic or a hyperbolic orbit, because such orbits are only appropriate to objects that are ultimately able to escape from the Sun's gravitational field.

14.7 KEPLER'S THIRD LAW

We have seen that the radius vector connecting our planet to the origin sweeps out area at the constant rate $dA/dt = h/2$. [See Equation (14.38).] We have also seen that the planetary orbit is an ellipse. Suppose that the major and minor radii of the ellipse are a and b, respectively. It follows that the area of the ellipse is $A = \pi a b$. Now, we expect the radius vector to sweep out the whole area of the ellipse in a single orbital period, T. Hence,

$$T = \frac{A}{(dA/dt)} = \frac{2\pi a b}{h}.$$

(14.46)

It follows from Equations (14.25), (14.26), and (14.45) that

$$T^2 = \frac{4\pi^2 a^3}{G M}.$$

(14.47)

In other words, the square of the orbital period of our planet is proportional to the cube of its orbital major radius. This is Kepler's third law of planetary motion.

14.8 ORBITAL PARAMETERS

Note that for an elliptical orbit the closest distance to the Sun—the so-called *perihelion* distance—is

$$r_p = \frac{r_c}{1+e} = a\,(1-e). \tag{14.48}$$

[See Equations (14.25) and (14.44).] Likewise, the furthest distance from the Sun—the so-called *aphelion* distance—is

$$r_a = \frac{r_c}{1-e} = a\,(1+e). \tag{14.49}$$

It follows that the major radius, a, is simply the mean of the perihelion and aphelion distances,

$$a = \frac{r_p + r_a}{2}. \tag{14.50}$$

The parameter

$$e = \frac{r_a - r_p}{r_a + r_p} \tag{14.51}$$

is called the *eccentricity*, and measures the deviation of the orbit from circularity. Thus, $e = 0$ corresponds to a circular orbit, whereas $e \to 1$ corresponds to an infinitely elongated elliptical orbit. The orbital parameters for the various planets in the solar system are listed in Table 14.1.

As is easily demonstrated from the previous analysis, Kepler laws of planetary motion can be written in the convenient form

$$r = \frac{a\,(1-e^2)}{1-e\cos\theta}, \tag{14.52}$$

$$r^2\,\dot\theta = (1-e^2)^{1/2}\,n\,a^2, \tag{14.53}$$

$$G\,M = n^2\,a^3, \tag{14.54}$$

where a is the mean orbital radius (i.e., the major radius), e the orbital eccentricity, and $n = 2\pi/T$ the mean orbital angular velocity.

14.9 ORBITAL ENERGIES

According to Equations (14.6), (14.9), and (14.10), the total energy per unit mass of an object in orbit around the Sun is given by

$$\mathcal{E} = \frac{\dot r^2 + r^2\dot\theta^2}{2} - \frac{G\,M}{r}. \tag{14.55}$$

It follows from Equations (14.36), (14.40), and (14.45) that

$$\mathcal{E} = \frac{h^2}{2}\left[\left(\frac{du}{d\theta}\right)^2 + u^2 - 2\,u\,u_c\right], \tag{14.56}$$

where $u = r^{-1}$, and $u_c = r_c^{-1}$. However, according to Equation (14.44),

$$u(\theta) = u_c\,(1 - e\cos\theta). \tag{14.57}$$

The previous two equations can be combined with Equations (14.45) and (14.48) to give

$$\mathcal{E} = \frac{u_c^2\,h^2}{2}\,(e^2 - 1) = \frac{G\,M}{2\,r_p}\,(e - 1). \tag{14.58}$$

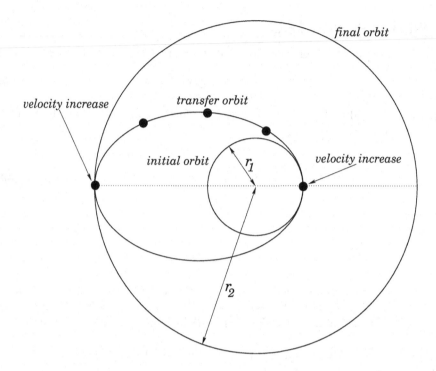

Figure 14.6 A transfer orbit between two circular orbits. (Reproduced from Fitzpatrick 2012. Courtesy of Cambridge University Press.)

We conclude that elliptical orbits ($e < 1$) have negative total energies, whereas parabolic orbits ($e = 1$) have zero total energies and hyperbolic orbits ($e > 1$) have positive total energies. This makes sense, because in a conservative system in which the potential energy at infinity is set to zero [see Equation (14.5)], we expect bounded orbits to have negative total energies, and unbounded orbits to have positive total energies. (See Section 5.7.) Thus, elliptical orbits, which are clearly bounded, should indeed have negative total energies, whereas hyperbolic orbits, which are clearly unbounded, should indeed have positive total energies. Parabolic orbits are marginally bounded (i.e., an object executing a parabolic orbit only just escapes from the Sun's gravitational field), and thus have zero total energy. For the special case of an elliptical orbit, whose major radius a is finite, we can write

$$\mathcal{E} = -\frac{G\,M}{2\,a}. \tag{14.59}$$

It follows that the energy of such an orbit is completely determined by the orbital major radius.

14.10   TRANSFER ORBITS

Consider an artificial satellite in an elliptical orbit around the Sun (the same considerations also apply to satellites in orbit around the Earth). At perihelion, $\dot{r} = 0$, and Equations (14.55) and (14.58) can be combined to give

$$\frac{v_t}{v_c} = \sqrt{1 + e}. \tag{14.60}$$

Here, $v_t = r\,\dot\theta$ is the satellite's tangential velocity and $v_c = (G\,M/r_p)^{1/2}$ is the tangential velocity that it would need in order to maintain a circular orbit at the perihelion distance. Likewise, at aphelion,

$$\frac{v_t}{v_c} = \sqrt{1 - e}, \qquad (14.61)$$

where $v_c = (G\,M/r_a)^{1/2}$ is now the tangential velocity that the satellite would need in order to maintain a circular orbit at the aphelion distance.

Suppose that our satellite is initially in a circular orbit of radius r_1 and that we wish to transfer it into a circular orbit of radius r_2, where $r_2 > r_1$. We can achieve this by temporarily placing the satellite in an elliptical orbit whose perihelion distance is r_1, and whose aphelion distance is r_2. It follows, from Equation (14.51), that the required eccentricity of the elliptical orbit is

$$e = \frac{r_2 - r_1}{r_2 + r_1}. \qquad (14.62)$$

According to Equation (14.60), we can transfer our satellite from its initial circular orbit into the temporary elliptical orbit by increasing its tangential velocity (by briefly switching on the satellite's rocket motor) by a factor

$$\alpha_1 = \sqrt{1 + e}. \qquad (14.63)$$

We must next allow the satellite to execute half an orbit, so that it attains its aphelion distance, and then boost the tangential velocity by a factor [see Equation (14.61)]

$$\alpha_2 = \frac{1}{\sqrt{1 - e}}. \qquad (14.64)$$

The satellite will now be in a circular orbit at the aphelion distance, r_2. This process is illustrated in Figure 14.6. Obviously, we can transfer our satellite from a larger to a smaller circular orbit by performing the previous process in reverse. Note, finally, from Equation (14.60), that if we increase the tangential velocity of a satellite in a circular orbit about the Sun by a factor greater than $\sqrt{2}$ then we will transfer it into a hyperbolic orbit ($e > 1$), and it will eventually escape from the Sun's gravitational field.

14.11 LOW-ECCENTRICITY ORBITS

It is clear from Table 14.1 that the planets in the solar system all have low-eccentricity (i.e., $e \ll 1$) elliptical orbits. It is possible to make a number of simplifying approximations for such orbits.

Consider a planet in a low-eccentricity elliptical orbit around the Sun. Suppose that the planet passes through its aphelion point, which corresponds to $\theta = 0$ at $t = 0$. Equations (14.52) and (14.53) can be combined to give

$$\frac{d(n\,t)}{d\theta} = \frac{(1 - e^2)^{3/2}}{(1 - e\,\cos\theta)^2}. \qquad (14.65)$$

Integration yields

$$n\,t = (1 - e^2)^{3/2} \int_0^\theta \frac{d\theta'}{(1 - e\,\cos\theta')^2}. \qquad (14.66)$$

Treating e as a small parameter, and expanding, we obtain

$$n t = \int_0^\theta \left(1 - \frac{3}{2}e^2 + \cdots\right)\left(1 + 2e\cos\theta' + 3e^2\cos^2\theta' + \cdots\right)d\theta'$$

$$= \int_0^\theta \left[1 + 2e\cos\theta' + \frac{3}{2}e^2\cos(2\theta) + \cdots\right]d\theta', \qquad (14.67)$$

which implies that

$$n t = \theta + 2e\sin\theta + \frac{3}{4}e^2\sin(2\theta) + \mathcal{O}(e^3). \qquad (14.68)$$

We can rearrange the previous expression to give

$$\theta = n t - 2e\sin\theta - \frac{3}{4}e^2\sin(2\theta) + \mathcal{O}(e^3). \qquad (14.69)$$

To zeroth order in e, we have $\theta = n t$. To first order in e, we have $\theta = n t - 2e\sin(n t)$. Finally, to second order in e, we have

$$\theta = n t - 2e\sin[n t - 2e\sin(n t)] - \frac{3}{4}e^2\sin(2 n t) + \mathcal{O}(e^3), \qquad (14.70)$$

which reduces to

$$\theta = n t - 2e\sin(n t) + \frac{5}{4}e^2\sin(2 n t) + \mathcal{O}(e^3). \qquad (14.71)$$

Finally,

$$\frac{r}{a} = \frac{1 - e^2}{1 - e\cos\theta} = (1 - e^2)\left[1 + e\cos\theta + e^2\cos^2\theta + \cdots\right]$$

$$= 1 + e\cos\theta - \frac{e^2}{2} + \frac{e^2}{2}\cos(2\theta) + \cdots$$

$$= 1 + e\cos[n t - 2e\sin(n t)] - \frac{e^2}{2} + \frac{e^2}{2}\cos(2 n t) + \cdots, \qquad (14.72)$$

which reduces to

$$\frac{r}{a} = 1 + e\cos(n t) + \frac{e^2}{2}\left[1 - \cos(2 n t)\right] + \mathcal{O}(e^3). \qquad (14.73)$$

Equations (14.71) and (14.73) effectively specify the position of the planet as a function of time.

14.12   TWO-BODY DYNAMICS

A two-body system is defined as an isolated dynamical system consisting of two objects that exert forces on one another. Suppose that our first object is of mass m_1 and is located at position vector $\mathbf{r}_1$. Likewise, our second object is of mass m_2 and is located at position vector $\mathbf{r}_2$. Let the first object exert a force $\mathbf{f}_{21}$ on the second. By Newton's third law, the second object exerts an equal and opposite force, $\mathbf{f}_{12} = -\mathbf{f}_{21}$, on the first. Suppose that there are no other forces in the problem. The equations of motion of our two objects are thus

$$m_1\frac{d^2\mathbf{r}_1}{dt^2} = -\mathbf{f}, \qquad (14.74)$$

$$m_2\frac{d^2\mathbf{r}_2}{dt^2} = \mathbf{f}, \qquad (14.75)$$

where $\mathbf{f} = \mathbf{f}_{21}$.

The center of mass of our system is located at

$$\mathbf{r}_{cm} = \frac{m_1\,\mathbf{r}_1 + m_2\,\mathbf{r}_2}{m_1 + m_2}. \tag{14.76}$$

Hence, we can write

$$\mathbf{r}_1 = \mathbf{r}_{cm} - \frac{m_2}{m_1 + m_2}\,\mathbf{r}, \tag{14.77}$$

$$\mathbf{r}_2 = \mathbf{r}_{cm} + \frac{m_1}{m_1 + m_2}\,\mathbf{r}, \tag{14.78}$$

where $\mathbf{r} = \mathbf{r}_2 - \mathbf{r}_1$. Substituting the previous two equations into Equations (14.74) and (14.75), and making use of the fact that the center of mass of an isolated system does not accelerate (see Section 6.3), we find that both equations yield

$$\mu\,\frac{d^2\mathbf{r}}{dt^2} = \mathbf{f}, \tag{14.79}$$

where

$$\mu = \frac{m_1\,m_2}{m_1 + m_2} \tag{14.80}$$

is called the *reduced mass*. Hence, we have effectively converted our original two-body problem into an equivalent one-body problem. In the equivalent problem, the force, $\mathbf{f}$, is the same as that acting on both objects in the original problem (modulo a minus sign). However, the mass, μ, is different, and is less than either of m_1 or m_2 (which is why it is called the "reduced" mass).

14.12.1   Binary Star Systems

Approximately half of the stars in our galaxy are members of so-called *binary star systems*. Such systems consist of two stars orbiting about their common center of mass. The distance separating the stars is always very much less than the distance to the nearest-neighbor star. Hence, a binary star system can be treated as a two-body dynamical system to a very good approximation.

In a binary star system, the gravitational force that the first star exerts on the second is

$$\mathbf{f} = -\frac{G\,m_1\,m_2}{r^3}\,\mathbf{r}, \tag{14.81}$$

where $\mathbf{r} = \mathbf{r}_2 - \mathbf{r}_1$. As we have seen, a two-body system can be reduced to an equivalent one-body system whose equation of motion is of the form (14.79), where $\mu = m_1\,m_2/(m_1 + m_2)$. Hence, in this particular case, we can write

$$\frac{m_1\,m_2}{m_1 + m_2}\,\frac{d^2\mathbf{r}}{dt^2} = -\frac{G\,m_1\,m_2}{r^3}\,\mathbf{r}, \tag{14.82}$$

which gives

$$\frac{d^2\mathbf{r}}{dt^2} = -\frac{G\,M}{r^3}\,\mathbf{r}, \tag{14.83}$$

where

$$M = m_1 + m_2. \tag{14.84}$$

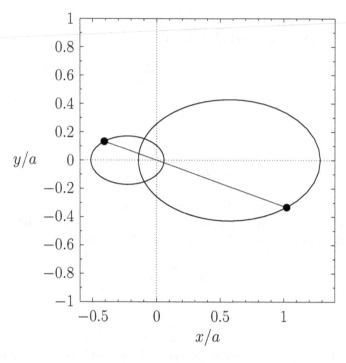

Figure 14.7 An example binary star orbit calculated with $m_1/m_2 = 0.4$ and $e = 0.8$. (Reproduced from Fitzpatrick 2012. Courtesy of Cambridge University Press.)

Equation (14.83) is identical to Equation (14.3), which we have already solved. Hence, we can immediately write down the solution:

$$\mathbf{r} = (r \cos\theta,\, r \sin\theta,\, 0), \tag{14.85}$$

where

$$r = \frac{a\,(1 - e^2)}{1 - e \cos\theta}, \tag{14.86}$$

and

$$\frac{d\theta}{dt} = \frac{h}{r^2}, \tag{14.87}$$

with

$$a = \frac{h^2}{(1 - e^2)\,G\,M}. \tag{14.88}$$

Here, h is a constant, and we have aligned our Cartesian axes so that the plane of the orbit coincides with the x-y plane. According to the previous solution, the second star executes a Keplerian elliptical orbit, with major radius a and eccentricity e, relative to the first star, and vice versa. From Equation (14.47), the period of revolution, T, is given by

$$T = \sqrt{\frac{4\pi^2\,a^3}{G\,M}}. \tag{14.89}$$

In the inertial frame of reference whose origin always coincides with the center of mass—the so-called center-of-mass frame—the position vectors of the two stars are

$$r_1 = -\frac{m_2}{m_1 + m_2}\, r, \tag{14.90}$$

$$r_2 = \frac{m_1}{m_1 + m_2}\, r, \tag{14.91}$$

where r was specified previously. Figure 14.7 shows an example binary star orbit, in the center-of-mass frame, calculated with $m_1/m_2 = 0.4$ and $e = 0.8$. It can be seen that both stars execute elliptical orbits about their common center of mass. Furthermore, at any given point in time, the stars are diagrammatically opposite one another, relative to the center of mass.

Binary star systems have been very useful to astronomers, because it is possible to determine the masses of both stars in such a system by careful observation. The sum of the masses of the two stars, $M = m_1 + m_2$, can be found from Equation (14.89) after a measurement of the major radius, a (which is the mean of the greatest and smallest distance apart of the two stars during their orbit), and the orbital period, T. The ratio of the masses of the two stars, m_1/m_2, can be determined from Equations (14.90) and (14.91) by observing the fixed ratio of the relative distances of the two stars from the common center of mass about which they both appear to rotate. Obviously, given the sum of the masses, and the ratio of the masses, the individual masses themselves can then be calculated.

14.13 EXERCISES

14.1 The distance of closest approach of Halley's comet to the Sun is 0.57 AU. (1 astronomical unit (AU) is the mean Earth-Sun distance.) The greatest distance of the comet from the Sun is 35 AU. The comet's speed at closest approach is 54 km/s. What is its speed when it is furthest from the Sun? [Ans: 0.879 km/s.]

14.2 A comet is observed a distance R astronomical units from the Sun, traveling at a speed that is V times the Earth's mean orbital speed.

   (a) Show that the orbit of the comet is hyperbolic, parabolic, or elliptical, depending on whether the quantity $V^2 R$ is greater than, equal to, or less than 2, respectively.

   (b) If the comet's observed direction of motion subtends an angle ϕ with the radius vector from the Sun, show that the comet's orbital eccentricity is

$$e = \left[1 + \left(V^2 - \frac{2}{R}\right)(R\,V\,\sin\phi)^2\right]^{1/2}.$$

14.3 If θ is the Sun's ecliptic longitude, measured from the perigee (i.e., the point of closest approach to the Earth), show that the Sun's apparent diameter is given by

$$D \simeq D_1 \cos^2\left(\frac{\theta}{2}\right) + D_2 \sin^2\left(\frac{\theta}{2}\right),$$

where D_1 and D_2 are the greatest and least values of D, respectively. [From Lamb 1942.]

14.4 A comet is in a parabolic orbit lying in the plane of the Earth's orbit, which is regarded as a circular orbit of radius a.

(a) Show that the points where the comet's orbit intersects the Earth's orbit are given by

$$\cos\theta = -1 + \frac{2p}{a},$$

where p is the perihelion distance of the comet defined at $\theta = 0$.

(b) Show that the time interval that the comet remains inside the Earth's orbit is the fraction

$$\frac{\sqrt{2}}{3\pi}\left(\frac{2p}{a}+1\right)\left(1-\frac{p}{a}\right)^{1/2}$$

of a year.

(c) Show that the maximum value of this time interval is $2/(3\pi)$ year, or 77.5 days, corresponding to $p = a/2$.

(d) Compute the time interval for Halley's comet ($p = 0.57\,a$). [Ans: 76.9 days.]

(Note: A high-eccentricity cometary orbit is well approximated as a parabolic orbit close to the perihelion point.)

14.5 The period and eccentricity of the Earth's orbit are $T = 365.24$ days and $e = 0.01673$, respectively.

(a) How long (in days) does it take the Sun-Earth radius vector to rotate through $90°$, starting at the perihelion point? [Ans: 89.4 days.]

(b) How long does it take starting at the aphelion point? [Ans: 93.3 days.]

14.6 If θ measures the Earth's angular position from its aphelion point then the spring (vernal) equinox, summer solstice, autumn equinox, and winter solstice occur when $\theta = 77.07°$, $77.07°+90°$, $77.07°+180°$, and $77.07°+270°$, respectively. The period and eccentricity of the Earth's orbit are $T = 365.24$ days and $e = 0.01673$, respectively.

(a) What is the length of spring (i.e., the time elapsed between a summer solstice and the preceding spring equinox)? [Ans: 92.76 days.]

(b) What is the length of summer (i.e., the time elapsed between an autumn equinox and the preceding summer solstice)? [Ans: 93.65 days.]

(c) What is the length of autumn (i.e., the time elapsed between a winter solstice and the preceding autumn equinox)? [Ans: 89.84 days.]

(d) What is the length of winter (i.e., the time elapsed between a spring equinox and the preceding winter solstice)? [Ans: 88.99 days.]

14.7 Show that the time-averaged apparent diameter of the Sun, as seen from a planet describing a low-eccentricity elliptical orbit, is approximately equal to the apparent diameter when the planet's distance from the Sun equals the major radius of the orbit. [From Lamb 1942.]

14.8 (a) Using the notation of Section 14.12, show that the total momentum and angular momentum of a two-body system take the form

$$\mathbf{P} = M\,\dot{\mathbf{r}}_{cm},$$
$$\mathbf{L} = M\,\mathbf{r}_{cm} \times \dot{\mathbf{r}}_{cm} + \mu\,\mathbf{r} \times \dot{\mathbf{r}},$$

respectively, where $M = m_1 + m_2$, and $\dot{} \equiv d/dt$.

(b) If the force acting between the bodies is conservative, such that $\mathbf{f} = -\nabla U$, demonstrate that the total energy of the system is written

$$E = \frac{1}{2} M \dot{r}_{cm}^2 + \frac{1}{2} \mu \dot{r}^2 + U.$$

Show, from the equation of motion, $\mu \ddot{\mathbf{r}} = -\nabla U$, that E is constant in time.

(c) If the force acting between the particles is central, so that $\mathbf{f} \propto \mathbf{r}$, demonstrate, from the equation of motion, $\mu \ddot{\mathbf{r}} = \mathbf{f}$, that $\mathbf{L}$ is constant in time.

CHAPTER 15

Gravitational Potential Theory

15.1 INTRODUCTION

The aim of this chapter is to employ **gravitational potential theory** to investigate the rotational flattening and tidal elongation of celestial bodies, as well as the precession of the equinoxes.

15.2 GRAVITATIONAL POTENTIAL

Consider two point masses, m and m', located at position vectors $\mathbf{r}$ and $\mathbf{r}'$, respectively. The vector gravitational force, $\mathbf{f}$, that mass m' exerts on mass m is written

$$\mathbf{f} = G\,m\,m'\,\frac{\mathbf{r}' - \mathbf{r}}{|\mathbf{r}' - \mathbf{r}|^3}. \tag{15.1}$$

Hence, the acceleration, $\mathbf{g}$, of mass m as a result of the gravitational force exerted on it by mass m' takes the form

$$\mathbf{g} = G\,m'\,\frac{\mathbf{r}' - \mathbf{r}}{|\mathbf{r}' - \mathbf{r}|^3}. \tag{15.2}$$

Now, the x-component of this acceleration is written

$$g_x = G\,m'\,\frac{x' - x}{[(x' - x)^2 + (y' - y)^2 + (z' - z)^2]^{3/2}}, \tag{15.3}$$

where $\mathbf{r} = (x,\,y,\,z)$ and $\mathbf{r}' = (x',\,y',\,z')$. However, as is easily demonstrated,

$$\frac{x' - x}{[(x' - x)^2 + (y' - y)^2 + (z' - z)^2]^{3/2}} \equiv \frac{\partial}{\partial x}\left(\frac{1}{[(x' - x)^2 + (y' - y)^2 + (z' - z)^2]^{1/2}}\right). \tag{15.4}$$

Hence,

$$g_x = G\,m'\,\frac{\partial}{\partial x}\left(\frac{1}{|\mathbf{r}' - \mathbf{r}|}\right), \tag{15.5}$$

with analogous expressions for g_y and g_z. It follows that

$$\mathbf{g} = -\nabla\Phi \equiv -\left(\frac{\partial\Phi}{\partial x},\,\frac{\partial\Phi}{\partial y},\,\frac{\partial\Phi}{\partial z}\right), \tag{15.6}$$

DOI: 10.1201/9781003198642-15

where

$$\Phi(\mathbf{r}) = -\frac{G\,m'}{|\mathbf{r}' - \mathbf{r}|} \tag{15.7}$$

is termed the *gravitational potential*. Incidentally, a comparison between Equation (5.44) and Equation (15.6) reveals that the gravitational potential energy of a mass m located at position vector $\mathbf{r}$ in a gravitational field whose gravitational potential is $\Phi(\mathbf{r})$ is simply $U = m\,\Phi(\mathbf{r})$.

It is a well known experimental fact that gravity is a superposable force. In other words, the gravitational force exerted on some point mass by a collection of other point masses is simply the vector sum of the forces exerted on the former mass by each of the latter masses taken in isolation. It follows that the gravitational potential generated by a collection of point masses at a certain location in space is the sum of the potentials generated at that location by each point mass taken in isolation. Hence, using Equation (15.7), if there are N point masses, m_i (for $i = 1, N$), located at position vectors $\mathbf{r}_i$, then the gravitational potential generated at position vector $\mathbf{r}$ is simply

$$\Phi(\mathbf{r}) = -G \sum_{i=1,N} \frac{m_i}{|\mathbf{r}_i - \mathbf{r}|}. \tag{15.8}$$

Suppose, finally, that, instead of having a collection of point masses, we have a continuous mass distribution. In other words, let the mass at position vector $\mathbf{r}'$ be $\rho(\mathbf{r}')\,d^3\mathbf{r}'$, where $\rho(\mathbf{r}')$ is the local mass density and $d^3\mathbf{r}'$ a volume element. Summing over all space, and taking the limit $d^3\mathbf{r}' \to 0$, we find that Equation (15.8) yields

$$\Phi(\mathbf{r}) = -G \int \frac{\rho(\mathbf{r}')}{|\mathbf{r}' - \mathbf{r}|}\,d^3\mathbf{r}', \tag{15.9}$$

where the integral is taken over all space. This is the general expression for the gravitational potential, $\Phi(\mathbf{r})$, generated by a continuous mass distribution, $\rho(\mathbf{r})$.

15.3 AXIALLY-SYMMETRIC MASS DISTRIBUTIONS

At this point, it is convenient to adopt standard spherical coordinates, r, θ, ϕ, aligned along the z-axis. These coordinates are related to regular Cartesian coordinates as follows:

$$x = r\sin\theta\,\cos\phi, \tag{15.10}$$

$$y = r\sin\theta\,\sin\phi, \tag{15.11}$$

$$z = r\cos\theta. \tag{15.12}$$

Consider an *axially-symmetric* mass distribution; that is, a $\rho(\mathbf{r})$ that is independent of the azimuthal angle, ϕ. We would expect such a mass distribution to generated an axially-symmetric gravitational potential, $\Phi(r, \theta)$. Hence, without loss of generality, we can set $\phi = 0$ when evaluating $\Phi(\mathbf{r})$ from Equation (15.9). In fact, given that $d^3\mathbf{r}' = r'^2\,\sin\theta'\,dr'\,d\theta'\,d\phi'$ in spherical coordinates, this equation yields

$$\Phi(r, \theta) = -G \int_0^\infty \int_0^\pi \int_0^{2\pi} \frac{r'^2\,\rho(r', \theta')\,\sin\theta'}{|\mathbf{r} - \mathbf{r}'|}\,d\phi'\,d\theta'\,dr', \tag{15.13}$$

with the right-hand side evaluated at $\phi = 0$. However, because $\rho(r', \theta')$ is independent of ϕ', Equation (15.13) can also be written

$$\Phi(r, \theta) = -2\pi\,G \int_0^\infty \int_0^\pi r'^2\,\rho(r', \theta')\,\sin\theta'\,\langle|\mathbf{r} - \mathbf{r}'|^{-1}\rangle\,d\theta'\,dr', \tag{15.14}$$

where $\langle \cdots \rangle \equiv \oint (\cdots) \, d\phi'/2\pi$ denotes an average over the azimuthal angle.

Now,

$$|\mathbf{r}' - \mathbf{r}|^{-1} = (r^2 - 2\,\mathbf{r} \cdot \mathbf{r}' + r'^2)^{-1/2}, \qquad (15.15)$$

and

$$\mathbf{r} \cdot \mathbf{r}' = r\,r'\,F, \qquad (15.16)$$

where (at $\phi = 0$)

$$F = \sin\theta \, \sin\theta' \, \cos\phi' + \cos\theta \, \cos\theta'. \qquad (15.17)$$

Hence,

$$|\mathbf{r}' - \mathbf{r}|^{-1} = (r^2 - 2\,r\,r'\,F + r'^2)^{-1/2}. \qquad (15.18)$$

Suppose that $r > r'$. In this case, we can expand $|\mathbf{r}' - \mathbf{r}|^{-1}$ as a convergent power series in r'/r, to give

$$|\mathbf{r}' - \mathbf{r}|^{-1} = \frac{1}{r}\left[1 + \left(\frac{r'}{r}\right) F + \frac{1}{2}\left(\frac{r'}{r}\right)^2 (3\,F^2 - 1) + \mathcal{O}\left(\frac{r'}{r}\right)^3 \right]. \qquad (15.19)$$

Let us now average this expression over the azimuthal angle, ϕ'. Because $\langle 1 \rangle = 1$, $\langle \cos\phi' \rangle = 0$, and $\langle \cos^2\phi' \rangle = 1/2$, it is easily seen that

$$\langle F \rangle = \cos\theta \, \cos\theta', \qquad (15.20)$$

$$\langle F^2 \rangle = \frac{1}{2}\,\sin^2\theta \, \sin^2\theta' + \cos^2\theta \, \cos^2\theta'$$
$$= \frac{1}{3} + \frac{2}{3}\left(\frac{3}{2}\cos^2\theta - \frac{1}{2}\right)\left(\frac{3}{2}\cos^2\theta' - \frac{1}{2}\right). \qquad (15.21)$$

Hence,

$$\langle |\mathbf{r}' - \mathbf{r}|^{-1} \rangle = \frac{1}{r}\left[1 + \left(\frac{r'}{r}\right) \cos\theta \, \cos\theta' \right.$$
$$\left. + \left(\frac{r'}{r}\right)^2 \left(\frac{3}{2}\cos^2\theta - \frac{1}{2}\right)\left(\frac{3}{2}\cos^2\theta' - \frac{1}{2}\right) + \mathcal{O}\left(\frac{r'}{r}\right)^3 \right]. \qquad (15.22)$$

Now, the well-known *Legendre polynomials*, $P_n(x)$, are defined as

$$P_n(x) = \frac{1}{2^n \, n!}\frac{d^n}{dx^n}\left[(x^2 - 1)^n\right], \qquad (15.23)$$

for $n = 0, \infty$. It follows that

$$P_0(x) = 1, \qquad (15.24)$$
$$P_1(x) = x, \qquad (15.25)$$
$$P_2(x) = \frac{1}{2}\,(3\,x^2 - 1), \qquad (15.26)$$

and so on. The Legendre polynomials are mutually orthogonal:

$$\int_{-1}^{1} P_n(x)\,P_m(x)\,dx = \int_0^{\pi} P_n(\cos\theta)\,P_m(\cos\theta)\,\sin\theta \, d\theta = \frac{\delta_{nm}}{n + 1/2}. \qquad (15.27)$$

Here, δ_{nm} is 1 if $n = m$, and 0 otherwise. The Legendre polynomials also form a complete set; that is, any function of x that is well behaved in the interval $-1 \le x \le 1$ can be represented as a weighted sum of the $P_n(x)$. Likewise, any function of θ that is well behaved in the interval $0 \le \theta \le \pi$ can be represented as a weighted sum of the $P_n(\cos\theta)$.

A comparison of Equation (15.22) and Equations (15.24)–(15.26) makes it reasonably clear that, when $r > r'$, the complete expansion of $\langle |\mathbf{r}' - \mathbf{r}|^{-1}\rangle$ is

$$\langle |\mathbf{r}' - \mathbf{r}|^{-1}\rangle = \frac{1}{r} \sum_{n=0,\infty} \left(\frac{r'}{r}\right)^n P_n(\cos\theta)\, P_n(\cos\theta'). \tag{15.28}$$

Similarly, when $r < r'$, we can expand in powers of r/r' to obtain

$$\langle |\mathbf{r}' - \mathbf{r}|^{-1}\rangle = \frac{1}{r'} \sum_{n=0,\infty} \left(\frac{r}{r'}\right)^n P_n(\cos\theta)\, P_n(\cos\theta'). \tag{15.29}$$

It follows from Equations (15.14), (15.28), and (15.29) that

$$\Phi(r,\theta) = \sum_{n=0,\infty} \Phi_n(r)\, P_n(\cos\theta), \tag{15.30}$$

where

$$\Phi_n(r) = -\frac{2\pi\, G}{r^{n+1}} \int_0^r \int_0^\pi r'^{\,n+2} \rho(r',\theta')\, P_n(\cos\theta')\, \sin\theta'\, d\theta'\, dr'$$
$$- 2\pi\, G\, r^n \int_r^\infty \int_0^\pi r'^{\,1-n} \rho(r',\theta')\, P_n(\cos\theta')\, \sin\theta'\, d\theta'\, dr'. \tag{15.31}$$

Given that the $P_n(\cos\theta)$ form a complete set, we can always write

$$\rho(r,\theta) = \sum_{n=0,\infty} \rho_n(r)\, P_n(\cos\theta). \tag{15.32}$$

This expression can be inverted, with the aid of Equation (15.27), to give

$$\rho_n(r) = (n + 1/2) \int_0^\pi \rho(r,\theta)\, P_n(\cos\theta)\, \sin\theta\, d\theta. \tag{15.33}$$

Hence, Equation (15.31) reduces to

$$\Phi_n(r) = -\frac{2\pi\, G}{(n+1/2)\, r^{n+1}} \int_0^r r'^{\,n+2} \rho_n(r')\, dr' - \frac{2\pi\, G\, r^n}{n+1/2} \int_r^\infty r'^{\,1-n} \rho_n(r')\, dr'. \tag{15.34}$$

Thus, we now have a general expression for the gravitational potential, $\Phi(r,\theta)$, generated by an axially-symmetric mass distribution, $\rho(r,\theta)$.

15.4   GRAVITATIONAL POTENTIAL DUE TO A UNIFORM SPHERE

Let us calculate the gravitational potential generated by a sphere of uniform mass density γ and radius R, whose center coincides with the origin. Expressing $\rho(r,\theta)$ in the form of Equation (15.32), we find that

$$\rho_0(r) = \begin{cases} \gamma & \text{for } r \le R \\ 0 & \text{for } r > R \end{cases}, \tag{15.35}$$

with $\rho_n(r) = 0$ for $n > 0$. Thus, from Equation (15.34),

$$\Phi_0(r) = -\frac{4\pi\, G\, \gamma}{r} \int_0^r r'^2\, dr' - 4\pi\, G\, \gamma \int_r^R r'\, dr' \tag{15.36}$$

for $r \leq R$, and

$$\Phi_0(r) = -\frac{4\pi\, G\, \gamma}{r} \int_0^R r'^2\, dr' \tag{15.37}$$

for $r > R$, with $\Phi_n(r) = 0$ for $n > 0$. Hence,

$$\Phi(r) = -\frac{2\pi\, G\, \gamma}{3}\, (3\,R^2 - r^2) = -G\,M\,\frac{(3\,R^2 - r^2)}{2\,R^3} \tag{15.38}$$

for $r \leq R$, and

$$\Phi(r) = -\frac{4\pi\, G\, \gamma}{3}\, \frac{R^3}{r} = -\frac{G\,M}{r} \tag{15.39}$$

for $r > R$. Here, $M = (4\pi/3)\, R^3\, \gamma$ is the total mass of the sphere.

According to Equation (15.39), the gravitational potential outside a uniform sphere of mass M is the same as that generated by a point mass, M, located at the sphere's center. It turns out that this is a general result for any finite spherically-symmetric mass distribution. Indeed, from the preceding analysis, it is clear that $\rho(r, \theta) = \rho_0(r)$ and $\Phi(r, \theta) = \Phi_0(r)$ for such a distribution. Suppose that the distribution extends out to $r = R$. It immediately follows, from Equation (15.34), that

$$\Phi_0(r) = -\frac{G}{r} \int_0^R 4\pi\, r'^2\, \rho_0(r')\, dr' = -\frac{G\,M}{r} \tag{15.40}$$

for $r > R$, where M is the total mass of the distribution. This, then, is the theoretical justification for Newton's famous result that the gravitational field generated outside a spherically-symmetric mass distribution is the same as that generated by an equivalent point mass located at the center of the distribution.

15.5 GRAVITATIONAL POTENTIAL OUTSIDE A UNIFORM SPHEROID

Let us now calculate the gravitational potential generated outside a spheroid of uniform mass density γ and mean radius R. A *spheroid* is the solid body produced by rotating an ellipse about a major or a minor axis. Let the axis of rotation coincide with the z-axis, and let the outer boundary of the spheroid satisfy

$$r = R_\theta(\theta) = R \left[1 - \frac{2}{3}\, \epsilon\, P_2(\cos\theta) \right], \tag{15.41}$$

where ϵ is termed the *ellipticity*. In fact, the radius of the spheroid at the poles (i.e., along the rotation axis) is $R_p = R\,(1 - 2\,\epsilon/3)$, whereas the radius at the equator (i.e., in the bisecting plane perpendicular to the axis) is $R_e = R\,(1 + \epsilon/3)$. Hence,

$$\epsilon = \frac{R_e - R_p}{R}. \tag{15.42}$$

Note that R is the surface-averaged radius: that is, $R = (1/4\pi) \oint \int_0^\pi R_\theta(\theta)\, \sin\theta\, d\theta\, d\phi$.

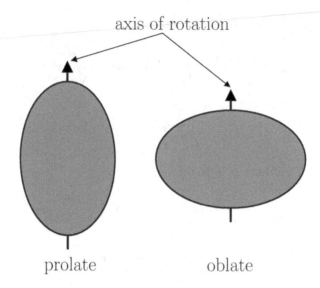

Figure 15.1 Prolate and oblate spheroids. (Reproduced from Fitzpatrick 2012. Courtesy of Cambridge University Press.)

Let us assume that $|\epsilon| \ll 1$, so that the spheroid is very close to being a sphere. If $\epsilon > 0$ then the spheroid is slightly squashed along its symmetry axis and is termed *oblate*. Likewise, if $\epsilon < 0$ then the spheroid is slightly elongated along its axis and is termed *prolate*. See Figure 15.1. Of course, if $\epsilon = 0$ then the spheroid reduces to a sphere.

Now, according to Equations (15.30) and (15.31), the gravitational potential generated outside an axially-symmetric mass distribution can be written

$$\Phi(r, \theta) = \frac{GM}{R} \sum_{n=0,\infty} J_n \left(\frac{R}{r} \right)^{n+1} P_n(\cos \theta), \tag{15.43}$$

where M is the total mass of the distribution, and

$$J_n = -\frac{2\pi R^3}{M} \int \int \left(\frac{r}{R} \right)^{2+n} \rho(r, \theta) \, P_n(\cos \theta) \, \sin \theta \, d\theta \, \frac{dr}{R}. \tag{15.44}$$

Here, the integral is taken over the whole cross-section of the distribution in r-θ space.

It follows that for a uniform spheroid, for which $M = (4\pi/3) \gamma R^3$,

$$J_n = -\frac{3}{2} \int_0^\pi P_n(\cos \theta) \int_0^{R_\theta(\theta)} \frac{r^{2+n} \, dr}{R^{3+n}} \sin \theta \, d\theta. \tag{15.45}$$

Hence,

$$J_n = -\frac{3}{2(3+n)} \int_0^\pi P_n(\cos \theta) \left[\frac{R_\theta(\theta)}{R} \right]^{3+n} \sin \theta \, d\theta, \tag{15.46}$$

giving

$$J_n \simeq -\frac{3}{2(3+n)} \int_0^\pi P_n(\cos \theta) \left[P_0(\cos \theta) - \frac{2}{3}(3+n) \, \epsilon \, P_2(\cos \theta) \right] \sin \theta \, d\theta, \tag{15.47}$$

to first order in ϵ. It is thus clear, from Equation (15.27), that, to first order in ϵ, the only nonzero J_n are

$$J_0 = -1, \tag{15.48}$$

$$J_2 = \frac{2}{5}\,\epsilon. \tag{15.49}$$

Thus, the gravitational potential outside a uniform spheroid of total mass M, mean radius R, and ellipticity ϵ is

$$\Phi(r,\theta) = -\frac{GM}{r} + \frac{2}{5}\,\epsilon\,\frac{GMR^2}{r^3}\,P_2(\cos\theta) + \mathcal{O}(\epsilon^2). \tag{15.50}$$

By analogy with the preceding analysis, the gravitational potential outside a general (i.e., axisymmetric, but not necessarily uniform) spheroidal mass distribution of mass M, mean radius R, and ellipticity ϵ (where $|\epsilon| \ll 1$) can be written

$$\Phi(r,\theta) = -\frac{GM}{r} + J_2\,\frac{GMR^2}{r^3}\,P_2(\cos\theta) + \mathcal{O}(\epsilon^2), \tag{15.51}$$

where $J_2 \sim \mathcal{O}(\epsilon)$.

15.6   ROTATIONAL FLATTENING

Consider the equilibrium configuration of a self-gravitating celestial body, composed of incompressible fluid, that is rotating steadily and uniformly at angular velocity Ω about some fixed axis passing through its center of mass.

Let us transform to a non-inertial frame of reference that co-rotates with the body about the z-axis, and in which the body consequently appears to be stationary. From Section 12.2, the problem is now analogous to that of a non-rotating body, except that the acceleration of a point mass located outside the distribution is written $\mathbf{g} = \mathbf{g}_g + \mathbf{g}_c$, where $\mathbf{g}_g = -\nabla\Phi(r,\theta)$ is the gravitational acceleration, $\mathbf{g}_c$ the centrifugal acceleration, and Φ the gravitational potential. The centrifugal acceleration is of magnitude $r\sin\theta\,\Omega^2$ and is everywhere directed away from the axis of rotation. (See Section 12.3.) Here, r and θ are spherical coordinates whose origin is the body's geometric center and whose symmetry axis coincides with the axis of rotation. The centrifugal acceleration is thus

$$\mathbf{g}_c = r\,\Omega^2\sin^2\theta\,\mathbf{e}_r + r\,\Omega^2\sin\theta\,\cos\theta\,\mathbf{e}_\theta. \tag{15.52}$$

It follows that $\mathbf{g}_c = -\nabla\chi$, where

$$\chi(r,\theta) = -\frac{\Omega^2 r^2}{2}\sin^2\theta = \frac{\Omega^2 r^2}{3}\,[P_2(\cos\theta) - 1] \tag{15.53}$$

can be thought of as a sort of centrifugal potential. Thus, the total acceleration is

$$\mathbf{g} = -\nabla(\Phi + \chi). \tag{15.54}$$

Let us assume that the outer boundary of the body is spheroidal. It follows that the boundary satisfies [see Equation (15.41)]

$$r = R_\theta(\theta) = R\left[1 - \frac{2}{3}\,\epsilon\,P_2(\cos\theta)\right], \tag{15.55}$$

where R is the body's mean (i.e., surface-averaged) radius, and ϵ its ellipticity. Obviously, we are choosing the symmetry axis of the spheroid to coincide with the axis of rotation (i.e., the z-axis). Note that $R_p = R\,(1 - 2\,\epsilon/3)$ the the body's *polar radius* (i.e., its radius along the axis of rotation), whereas $R_e = R\,(1 + \epsilon/3)$ is the body's *equatorial radius* (i.e., its radius in the bisecting plane that is perpendicular to the axis of rotation). Moreover,

$$\epsilon = \frac{R_e - R_p}{R}. \tag{15.56}$$

It is convenient to write the centrifugal potential in the form

$$\chi(r, \theta) = \frac{G\,M}{R}\,\zeta\left(\frac{r}{R}\right)^2 [P_2(\cos\theta) - 1], \tag{15.57}$$

where the dimensionless parameter

$$\zeta = \frac{\Omega^2\,R^3}{3\,G\,M} \tag{15.58}$$

is the typical ratio of the centrifugal acceleration to the gravitational acceleration at $r \simeq R$. Let us assume that both ϵ and ζ are small compared to unity. In other words, the body is almost spherical, and the typical centrifugal potential is much smaller than the typical gravitational acceleration.

Now, the condition for an equilibrium state is that the total potential is uniform over the body's surface. If this is not the case then, according to Equation (15.54), there will be net forces acting tangential to the surface. Such forces cannot be balanced by internal fluid pressure, which only acts normal to the surface. The external (to the body) gravitational potential can be written [see Equation (15.51)]

$$\Phi(r, \theta) \simeq -\frac{G\,M}{r} + J_2\,\frac{G\,M\,R^2}{r^3}\,P_2(\cos\theta), \tag{15.59}$$

where $J_2 \sim \mathcal{O}(\epsilon)$. The equilibrium configuration is specified by

$$\Phi(R_\theta, \theta) + \chi(R_\theta, \theta) = c, \tag{15.60}$$

where c is a constant. It follows from Equations (15.57), (15.55), and (15.59) that, to first order in ϵ and ζ,

$$-\frac{G\,M}{R}\left[1 + \left(\frac{2}{3}\epsilon - J_2\right) P_2(\cos\theta)\right] + \frac{G\,M}{R}\,\zeta\,[P_2(\cos\theta) - 1] \simeq c, \tag{15.61}$$

which yields

$$\epsilon = \frac{3}{2}\,(J_2 + \zeta). \tag{15.62}$$

For the special case of a uniform-density body, we have $J_2 = (2/5)\,\epsilon$. [See Equation (15.49).] Hence, the previous equation simplifies to

$$\epsilon = \frac{15}{4}\,\zeta, \tag{15.63}$$

or

$$\frac{R_e - R_p}{R} = \frac{5}{4}\,\frac{\Omega^2\,R^3}{G\,M}. \tag{15.64}$$

We conclude, from the preceding analysis, that the equilibrium configuration of a (relatively slowly) rotating self-gravitating fluid mass is an *oblate spheroid*; in other words, a sphere that is slightly flattened along its axis of rotation. The degree of flattening is proportional to the square of the rotation rate. Note that expression (15.64) first appeared (in a somewhat modified form) in Newton's *Principia*.

Equation (15.62) was derived for a rotating, self-gravitating, celestial body composed of incompressible fluid. Such a body has zero shear stress at its surface because fluids (by definition) are unable to withstand shear stresses. Solids, on the other hand, can withstand such stresses to a limited extent. Hence, it is not necessarily true that there is zero shear stress at the surface of a solid rotating celestial body, such as the Earth. However, it turns out that, for a solid rotating celestial body whose radius exceeds a few hundred kilometers, the shear stresses that would develop within the body, were it to attempt to resist the centrifugal acceleration generated by its rotation, would be so great that the rock composing the body would flow like a liquid. In other words, it is a reasonably good approximation to treat a rotating celestial body whose radius exceeds a few hundred kilometers as an incompressible fluid, irrespective of its internal makeup.

15.6.1 Rotational Flattening of Earth

For the case of the Earth, $R = 6.371 \times 10^6$ m, $\Omega = 7.2921 \times 10^{-5}$ rad/s, and $M = 5.972 \times 10^{24}$ kg (Yoder 1995). It follows that

$$\zeta = 1.15 \times 10^{-3}. \tag{15.65}$$

Assuming that the Earth is a uniform-density body, its rotation flattening is governed by Equation (15.63), which yields

$$\epsilon = 4.31 \times 10^{-3}. \tag{15.66}$$

This corresponds to a difference between the Earth's equatorial and polar radii of

$$\Delta R = R_e - R_p = \epsilon R = 27.5 \, \text{km}. \tag{15.67}$$

In fact, the observed degree of rotational flattening of the Earth is $\epsilon = 3.35 \times 10^{-3}$, corresponding to a difference between equatorial and polar radii of 21.4 km. Our analysis has overestimated the Earth's rotational flattening because we assumed that the terrestrial interior is of uniform density. In reality, the Earth's core is significantly denser than its crust. Incidentally, the observed value of the dimensionless parameter J_2, which measures the strength of the Earth's quadrupole gravitational field, is 1.08×10^{-3} (Yoder 1995). Hence, $(3/2)(J_2 + \zeta) = 3.35 \times 10^{-3}$. In other words, the Earth's rotational flattening satisfies Equation (15.62) extremely accurately. This confirms that, although the Earth is not a uniform-density body, the Earth's response to the centrifugal acceleration is indeed fluid-like [because Equation (15.62) was derived on the assumption that the surface of the rotating body in question is in hydrostatic equilibrium in the co-rotating frame].

15.7 TIDAL ELONGATION

Consider two point masses, m and m', executing circular orbits about their common center of mass, C, with angular velocity ω. Let a be the distance between the masses and ρ the distance between point C and mass m. See Figure 15.2. We know from Section 14.12.1 that

$$\omega^2 = \frac{G M}{a^3}, \tag{15.68}$$

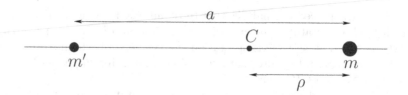

Figure 15.2 Two orbiting masses. (Reproduced from Fitzpatrick 2012. Courtesy of Cambridge University Press.)

and

$$\rho = \frac{m'}{M}\,a, \tag{15.69}$$

where $M = m + m'$.

Let us transform to a non-inertial frame of reference that rotates, about an axis perpendicular to the orbital plane and passing through C, at the angular velocity ω. In this reference frame, both masses appear to be stationary. Consider mass m. In the rotating frame, this mass experiences a gravitational acceleration

$$a_g = \frac{G\,m'}{a^2} \tag{15.70}$$

directed toward the center of mass, and a centrifugal acceleration (see Section 12.3)

$$a_c = \omega^2 \rho \tag{15.71}$$

directed away from the center of mass. However, it is easily demonstrated, using Equations (15.68) and (15.69), that

$$a_c = a_g. \tag{15.72}$$

In other words, the gravitational and centrifugal accelerations balance, as must be the case if mass m is to remain stationary in the rotating frame. Let us investigate how this balance is affected if the masses m and m' have finite spatial extents.

Let the center of the mass distribution m' lie at A, the center of the mass distribution m at B, and the center of mass at C. See Figure 15.3. We wish to calculate the centrifugal and gravitational accelerations at some point D in the vicinity of point B. It is convenient to adopt spherical coordinates, centered on point B, and aligned such that the z-axis coincides with the line BA.

Let us assume that the mass distribution m is orbiting around C, but is not rotating about an axis passing through its center of mass, in order to exclude rotational flattening

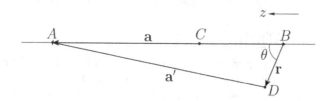

Figure 15.3 Calculation of tidal forces. (Reproduced from Fitzpatrick 2012. Courtesy of Cambridge University Press.)

from our analysis. If this is the case then it is easily seen that each constituent point of m executes circular motion of angular velocity ω and radius ρ. See Figure 15.4. Hence, each point experiences the same centrifugal acceleration:

$$\mathbf{g}_c = -\omega^2 \rho \, \mathbf{e}_z. \tag{15.73}$$

It follows that

$$\mathbf{g}_c = -\nabla \chi', \tag{15.74}$$

where

$$\chi' = \omega^2 \rho \, z \tag{15.75}$$

is the centrifugal potential and $z = r \cos \theta$. The centrifugal potential can also be written

$$\chi' = \frac{G \, m'}{a} \frac{r}{a} P_1(\cos \theta). \tag{15.76}$$

The gravitational acceleration at point D due to mass m' is given by

$$\mathbf{g}_g = -\nabla \Phi', \tag{15.77}$$

where the gravitational potential takes the form

$$\Phi' = -\frac{G \, m'}{a'}. \tag{15.78}$$

Here, a' is the distance between points A and D. The gravitational potential generated by the mass distribution m' is the same as that generated by an equivalent point mass at A, as long as the distribution is spherically symmetric, which we shall assume to be the case. Now,

$$\mathbf{a}' = \mathbf{a} - \mathbf{r}, \tag{15.79}$$

where $\mathbf{a}'$ is the vector $\overrightarrow{DA}$, and $\mathbf{a}$ the vector $\overrightarrow{BA}$. See Figure 15.3. It follows that

$$a'^{-1} = \left(a^2 - 2\, \mathbf{a} \cdot \mathbf{r} + r^2 \right)^{-1/2} = \left(a^2 - 2\, a \, r \cos \theta + r^2 \right)^{-1/2}. \tag{15.80}$$

Expanding in powers of r/a, we obtain

$$a'^{-1} = a^{-1} \sum_{n=0,\infty} \left(\frac{r}{a} \right)^n P_n(\cos \theta). \tag{15.81}$$

Hence,

$$\Phi' \simeq -\frac{G \, m'}{a} \left[1 + \frac{r}{a} P_1(\cos \theta) + \frac{r^2}{a^2} P_2(\cos \theta) \right], \tag{15.82}$$

to second order in r/a, where the $P_n(x)$ are Legendre polynomials.

Adding χ' and Φ', we find that

$$\chi = \chi' + \Phi' \simeq -\frac{G \, m'}{a} \left[1 + \frac{r^2}{a^2} P_2(\cos \theta) \right], \tag{15.83}$$

to second order in r/a. Note that χ is the potential due to the net externally generated force acting on the mass distribution m. This potential is constant up to first order in r/a, because the first-order variations in χ' and Φ' cancel each other. The cancelation is a manifestation of the balance between the centrifugal and gravitational accelerations in the equivalent

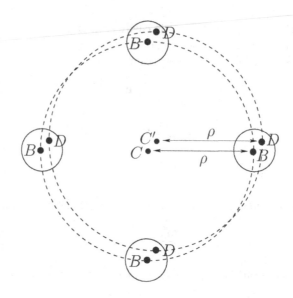

Figure 15.4 The center B of mass distribution m orbits about the center of mass C in a circle of radius ρ. If m is non-rotating then a noncentral point D maintains a constant spatial relationship to B, such that D orbits some point C' that has the same spatial relationship to C that D has to B, in a circle of radius ρ. (Reproduced from Fitzpatrick 2012. Courtesy of Cambridge University Press.)

point mass problem discussed previously. However, this balance is only exact at the center of the mass distribution m. Away from the center, the centrifugal acceleration remains constant, whereas the gravitational acceleration increases with increasing z. At positive z, the gravitational acceleration is larger than the centrifugal acceleration, giving rise to a net acceleration in the $+z$-direction. Likewise, at negative z, the centrifugal acceleration is larger than the gravitational, giving rise to a net acceleration in the $-z$-direction. It follows that the mass distribution m is subject to a residual acceleration, represented by the second-order variation in Equation (15.83), that acts to elongate it along the z-axis. This effect is known as *tidal elongation*.

Suppose that the mass distribution m is a uniform-density, liquid sphere of radius R, and uniform density γ. Let us estimate the elongation of this distribution due to the *tidal potential* specified in Equation (15.83), which (neglecting constant terms) can be written

$$\chi(r,\theta) = \frac{Gm}{R}\,\zeta\left(\frac{r}{R}\right)^2 P_2(\cos\theta). \tag{15.84}$$

Here, the dimensionless parameter

$$\zeta = -\frac{m'}{m}\left(\frac{R}{a}\right)^3 \tag{15.85}$$

is (minus) the typical ratio of the tidal acceleration to the gravitational acceleration at $r \simeq R$. Let us assume that $|\zeta| \ll 1$. By analogy with the analysis in the previous section, in the presence of the tidal potential the distribution becomes slightly spheroidal in shape, such that its outer boundary satisfies Equation (15.55). Moreover, the induced ellipticity, ϵ, of the distribution is related to the normalized amplitude, ζ, of the tidal potential according

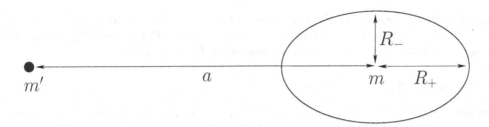

Figure 15.5 Tidal elongation. (Reproduced from Fitzpatrick 2012. Courtesy of Cambridge University Press.)

to Equation (15.63),

$$\epsilon = \frac{15}{4}\zeta. \tag{15.86}$$

15.7.1 Tidal Elongation of Earth Due to Moon

Consider the tidal elongation of the Earth due to the Moon. In this case, we have $R = 6.371 \times 10^6$ m, $a = 3.844 \times 10^8$ m, $m = 5.972 \times 10^{24}$ kg, and $m' = 7.342 \times 10^{22}$ kg (Yoder 1995). Hence, we find that

$$\zeta = -5.60 \times 10^{-8}. \tag{15.87}$$

According to Equation (15.86), the ellipticity of the Earth induced by the tidal effect of the Moon is

$$\epsilon = \frac{15}{4}\zeta \simeq -2.1 \times 10^{-7}. \tag{15.88}$$

The fact that ϵ is negative implies that the Earth is elongated along the z-axis; that is, along the axis joining its center to that of the Moon. [See Equation (15.55).] If R_+ and R_- are the greatest and least radii of the Earth, respectively, due to this elongation (see Figure 15.5), then

$$\Delta R = R_+ - R_- = -\epsilon R = 1.34 \, \text{m}. \tag{15.89}$$

Thus, we predict that the tidal effect of the Moon (which is actually due to spatial gradients in the Moon's gravitational field) causes the Earth to elongate along the axis joining its center to that of the Moon by 134 cm. This is an overestimate because the tidal potential is sufficiently weak that the Earth responds to it as an elastic solid, rather than as a fluid. Consequently, the Earth's actual tidal elongation is only about a quarter of the elongation of a hypothetical liquid Earth.

15.7.2 Tidal Elongation of Earth Due to Sun

Consider the tidal elongation of the Earth due to the Sun. In this case, we have $R = 6.371 \times 10^6$ m, $a = 1.496 \times 10^{11}$ m, $m = 5.972 \times 10^{24}$ kg, and $m' = 1.989 \times 10^{30}$ kg (Yoder 1995). Hence, we calculate that

$$\zeta = -2.57 \times 10^{-8}, \tag{15.90}$$

and

$$\epsilon = \frac{15}{4}\zeta = -9.6 \times 10^{-8}, \tag{15.91}$$

which implies that

$$\Delta R = R_+ - R_- = -\epsilon R = 0.61 \, \text{m}. \tag{15.92}$$

Thus, the tidal elongation of the Earth due to the Sun is 61 cm, which is about half that due to the Moon. As before, the true tidal elongation of the Earth due to the Sun is about one quarter of the previous estimate, because of the Earth's rigidity.

15.7.3 Ocean Tides

Because the Earth's oceans are liquid, their tidal elongation is significantly larger than that of the underlying land. (See Exercise 15.5.) Hence, the oceans rise, relative to the land, in the region of the Earth closest to the Moon, and also in the region furthest away. Because the Earth is rotating, while the tidal bulge of the oceans remains relatively stationary, the Moon's tidal effect causes the ocean at a given point on the Earth's surface to rise and fall twice daily, giving rise to the phenomenon known as the *tides*. There is also an oceanic tidal bulge due to the Sun that is about half as large as that due to the Moon. Consequently, ocean tides are particularly high when the Sun, the Earth, and the Moon lie approximately in a straight-line, so that the tidal effects of the Sun and the Moon reinforce one another. This occurs at a new moon, or at a full moon. These type of tides are called *spring tides* (the name has nothing to do with the season). Conversely, ocean tides are particularly low when the Sun, the Earth, and the Moon form a right angle, so that the tidal effects of the Sun and the Moon partially cancel one another. These type of tides are called *neap tides*. Generally speaking, we would expect two spring tides and two neap tides per month.

15.8 LUNI-SOLAR PRECESSION

Let us investigate the influence of the Sun on the Earth's diurnal rotation. Consider the Earth-Sun system. See Figure 15.6. From a geocentric viewpoint, the Sun orbits the Earth counter-clockwise (if we look from the north), once per year, in an approximately circular orbit of radius $a_s = 1.496 \times 10^{11}$ m (Yoder 1995). In astronomy, the plane of the Sun's apparent orbit relative to the Earth is known as the *ecliptic plane*. Let us define non-rotating Cartesian coordinates, centered on the Earth, which are such that the x- and y-axes lie in the ecliptic plane, and the z-axis is normal to this plane (in the sense that the Earth's north pole lies at positive z). It follows that the z-axis is directed toward a point in the sky (located in the constellation Draco) known as the *north ecliptic pole*. See Figure 15.7. In the following, we shall treat the x, y, z coordinate system as inertial. This is a reasonable approximation because the orbital acceleration of the Earth is much smaller than the acceleration due to its diurnal rotation. Let us parameterize the instantaneous position of the Sun in terms of a counter-clockwise (if we look from the north) azimuthal angle λ_s that is zero on the positive x-axis. See Figure 15.6.

Let $\boldsymbol{\omega}$ be the Earth's angular velocity vector due to its diurnal rotation. This vector subtends an angle θ with the z-axis, where $\theta = 23.44°$ is the mean inclination of the ecliptic to the Earth's equatorial plane (Yoder 1995). Suppose that the projection of $\boldsymbol{\omega}$ onto the ecliptic plane subtends an angle ϕ with the x-axis, where ϕ is measured in a counter-clockwise (if we look from the north) sense. See Figure 15.6. The orientation of the Earth's axis of rotation (which is, of course, parallel to $\boldsymbol{\omega}$) is thus determined by the two angles θ and ϕ.

According to Equation (15.51), assuming that the Sun is spherically symmetric, the gravitational potential energy of the Earth-Sun system is written

$$U_{es} = M_s \, \Phi = -\frac{G \, M_s \, M}{a_s} + J_2 \, \frac{G \, M_s \, M \, R^2}{a_s^3} \, P_2(\cos \gamma_s), \qquad (15.93)$$

where M_s is the mass of the Sun, M the mass of the Earth, R the mean radius of the Earth, and $J_2 = 1.08 \times 10^{-3}$ is a dimensionless measure of the Earth's quadrupole gravitational

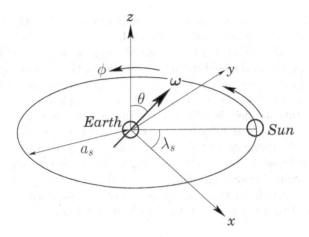

Figure 15.6 The Earth-Sun system. (Reproduced from Fitzpatrick 2012. Courtesy of Cambridge University Press.)

field generated by its rotational flattening. Furthermore, γ_s is the angle subtended between $\boldsymbol{\omega}$ and $\mathbf{r}_s$, where $\mathbf{r}_s$ is the position vector of the Sun relative to the Earth.

It is easily demonstrated that

$$\boldsymbol{\omega} = \omega\, (\sin\theta\,\cos\phi,\, \sin\theta\,\sin\phi,\, \cos\theta), \tag{15.94}$$

and

$$\mathbf{r}_s = a_s\, (\cos\lambda_s,\, \sin\lambda_s,\, 0). \tag{15.95}$$

Hence,

$$\cos\gamma_s = \frac{\boldsymbol{\omega}\cdot\mathbf{r}_s}{|\boldsymbol{\omega}|\,|\mathbf{r}_s|} = \sin\theta\,(\cos\phi\,\cos\lambda_s + \sin\phi\,\sin\lambda_s), \tag{15.96}$$

giving

$$U_{es} = -\frac{G\,M_s\,M}{a_s} + J_2\,\frac{G\,M_s\,M\,R^2}{2\,a_s^3}$$
$$\times\left[3\,\sin^2\theta\,(\cos^2\phi\,\cos^2\lambda_s + \sin^2\phi\,\sin^2\lambda_s + 2\,\cos\phi\,\sin\phi\,\cos\lambda_s\,\sin\lambda_s) - 1\right]. \tag{15.97}$$

Because we are primarily interested in the motion of the Earth's axis of rotation on timescales that are much longer than a year, we can average the preceding expression over the Sun's orbit to give

$$U_{es} = -\frac{G\,M_s\,M}{a_s} + J_2\,\frac{G\,M_s\,M\,R^2}{2\,a_s^3}\left(\frac{3}{2}\,\sin^2\theta - 1\right) \tag{15.98}$$

(because the average of $\cos^2\lambda_s$ and $\sin^2\lambda_s$ over a year is $1/2$, whereas the average of $\cos\lambda_s\,\sin\lambda_s$ is 0). Thus, we obtain

$$U_{es} = U_{es}^{(0)} - \frac{3}{8}\,J_2\,n_s^2\,M\,R^2\,\cos(2\,\theta), \tag{15.99}$$

where $U_{es}^{(0)}$ is a constant, and

$$n_s = \frac{d\lambda_s}{dt} = \left(\frac{G\,M_s}{a_s^3}\right)^{1/2} \tag{15.100}$$

is the Sun's apparent orbital angular velocity (which, of course, is really the Earth's orbital angular velocity around the Sun). (See Section 13.4.)

Now, we saw in Section 5.6 that if the potential energy of a particle (which is really the energy of the conservative force field within which the particle is moving) is a function of the Cartesian coordinate x—in other words, if $U = U(x)$—then the particle is subject to a force $f = -dU/dx$ acting in the x-direction. Hence, it is plausible that if the energy of the Earth-Sun gravitational field is a function of the angular coordinate θ then the Earth, which is moving in this field, is subject to a gravitational torque $-dU_{es}/d\theta$ acting in the θ direction; that is, the direction of the angular velocity imparted to the Earth were θ to increase in time. See Equation (9.57). (What we are really saying is that the decrease in the energy of the Earth-Sun gravitational field that would result if θ were to increase is due to the work done on the Earth, by the field, by means of the gravitational torque.) It follows that the gravitational torque exerted on the Earth by the Sun is

$$\boldsymbol{\tau} = -\frac{dU_{es}}{d\theta}\,(-\sin\phi,\,\cos\phi,\,0) = -\frac{3}{4}\,J_2\,n_s^2\,M\,R^2\,\sin(2\,\theta)\,(-\sin\phi,\,\cos\phi,\,0). \qquad (15.101)$$

See Section 9.6. The fact that the torque is proportional to J_2 suggests that the Sun can only exert a gravitational torque on the Earth because the Earth is rotationally flattened. The Earth's angular momentum is

$$\mathbf{L} = I_{\parallel}\,\boldsymbol{\omega} = I_{\parallel}\,\omega\,(\sin\theta\,\cos\phi,\,\sin\theta\,\sin\phi,\,\cos\theta), \qquad (15.102)$$

where $I_{\parallel}$ is the Earth's moment of inertia about its axis of rotation (which is a principal axis of rotation). (See Section 9.3.) The Earth's rotational equation of motion is (see Section 9.4)

$$\frac{d\mathbf{L}}{dt} = \boldsymbol{\tau}. \qquad (15.103)$$

The previous three equations can be satisfied if (see Section 9.6)

$$\frac{d\phi}{dt} = -\Omega_\phi, \qquad (15.104)$$

where

$$\Omega_\phi = \frac{3}{2}\,\frac{\tilde{\epsilon}\,n_s^2}{\omega}\,\cos\theta, \qquad (15.105)$$

and the dimensionless constant

$$\tilde{\epsilon} = J_2\,\frac{M\,R^2}{I_{\parallel}} = 3.27 \times 10^{-3}. \qquad (15.106)$$

is known as the Earth's *dynamical ellipticity*.

Note that if the Earth was a uniform body then $J_2 = (2/5)\,\epsilon$ [see Equation (15.49)] and $I_{\parallel} = (2/5)\,M\,R^2$ (see Section 8.5.4), which gives $\tilde{\epsilon} = \epsilon$. In other words, if the Earth was a uniform body then its dynamic ellipticity would exactly equal its geometric ellipticity, $\epsilon = \Delta R/R = 3.35 \times 10^{-3}$, where ΔR is the difference between the Earth's equatorial and polar radii. In reality, there is a slight difference between the Earth's dynamic and geometric ellipticities, due, in part, to the fact that the Earth's core is denser than its mantle. In fact, given that $J_2 = 1.08 \times 10^{-3}$ and $\tilde{\epsilon} = 3.27 \times 10^{-3}$, it is clear from Equation (15.106) that $I_{\parallel} = 0.330\,M\,R^2$. In other words, the moment of inertia of the Earth is less than that expected for a uniform Earth (i.e., $0.400\,M\,R^2$) as a consequence of the real Earth's central condensation.

According to the preceding equations, the torque that the Sun exerts on the Earth, as a consequence of its slight oblateness, causes the Earth's axis of rotation to precess steadily about the normal to the ecliptic plane at the rate $-\Omega_\phi$. The fact that $-\Omega_\phi$ is negative implies that the precession is in the opposite sense to that of the Earth's diurnal rotation and the Sun's apparent orbit about the Earth. The precession period in (sidereal) years is given by

$$T_\phi(\text{yr}) = \frac{n_s}{\Omega_\phi} = \frac{2\,T_s(\text{day})}{3\,\tilde{\epsilon}\,\cos\theta}, \tag{15.107}$$

where $T_s(\text{day}) = \omega/n_s = 366.26$ is the length of a sidereal year in sidereal days. (A sidereal year is the time required for the Sun to make a complete rotation in the sky with respect to the fixed stars. A sidereal day is the time required for a fixed star to make a complete rotation around the Earth.) Thus, given that $\tilde{\epsilon} = 3.27 \times 10^{-3}$ and $\theta = 23.44°$, we obtain

$$T_\phi \simeq 81\,400 \text{ years.} \tag{15.108}$$

Unfortunately, the observed precession period of the Earth's axis of rotation about the normal to the ecliptic plane is approximately 25 800 years, so something is clearly missing from our model. It turns out that the missing factor is the influence of the Moon.

The Moon orbits the Earth in an approximately circular orbit of radius $a_m = 3.844 \times 10^8$ m (Yoder 1995). If we neglect the slight inclination (5°) of the Moon's orbital plane to the ecliptic plane then the position vector of the Moon relative to the Earth can be written

$$\mathbf{r}_m = a_m\,(\cos\lambda_m, \sin\lambda_m, 0), \tag{15.109}$$

where λ_m is the angle subtended between the Earth-Moon vector and the x-axis. By analogy with Equation (15.93), assuming that the Moon is spherically symmetric, the gravitational potential energy of the Earth-Moon system is written

$$U_{em} = -\frac{G\,M_m\,M}{a_m} + J_2\,\frac{G\,M_m\,M\,R^2}{a_m^3}\,P_2(\cos\gamma_m), \tag{15.110}$$

where M_m is the mass of the Moon, and γ_m is the angle subtended between $\boldsymbol{\omega}$ and $\mathbf{r}_m$. It follows that

$$U_{em} = -\frac{G\,M_m\,M}{a_m} + J_2\,\frac{G\,M_m\,M\,R^2}{2\,a_m^3} \tag{15.111}$$
$$\times \left[3\,\sin^2\theta\,(\cos^2\phi\,\cos^2\lambda_m + \sin^2\phi\,\sin^2\lambda_m + 2\,\cos\phi\,\sin\phi\,\cos\lambda_s\,\sin\lambda_m) - 1\right].$$

Because we are primarily interested in the motion of the Earth's axis of rotation on timescales that are much longer than a month, we can average the preceding expression over the Moon's orbit to give

$$U_{em} = -\frac{G\,M_m\,M}{a_m} + J_2\,\frac{G\,M_m\,M\,R^2}{2\,a_m^3}\left(\frac{3}{2}\,\sin^2\theta - 1\right) \tag{15.112}$$

(because the average of $\cos^2\lambda_m$ and $\sin^2\lambda_m$ over a month is 1/2, whereas the average of $\cos\lambda_m\,\sin\lambda_m$ is 0). Thus, we obtain

$$U_{em} = U_{em}^{(0)} - \frac{3}{8}\,J_2\,\mu_m\,n_m^2\,M\,R^2\,\cos(2\,\theta), \tag{15.113}$$

where $U_{em}^{(0)}$ is a constant,

$$\mu_m = \frac{M_m}{M} \tag{15.114}$$

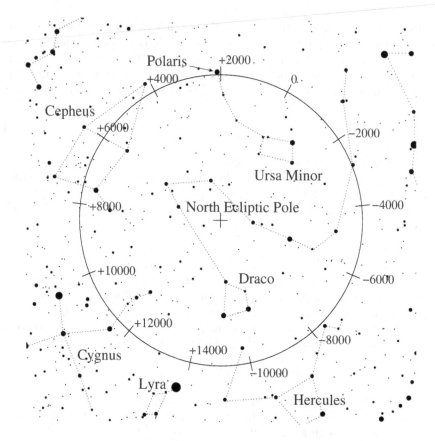

Figure 15.7 Path of the north celestial pole against the backdrop of the stars as consequence of the precession of the equinoxes (calculated assuming constant precessional speed and obliquity). Numbers indicate years relative to start of common era. (Reproduced from Fitzpatrick 2012. Courtesy of Cambridge University Press.)

is the ratio of the mass of the Moon to that of the Earth, and

$$n_m = \frac{d\lambda_m}{dt} = \left(\frac{GM}{a_m^3}\right)^{1/2} \tag{15.115}$$

is the Moon's orbital angular velocity. (See Section 13.4.)

Combining Equations (15.99) and (15.113), the gravitational potential energy of the Earth-Sun-Moon system can be written

$$U_{esm} = U_{es} + U_{em} = U_{esm}^{(0)} - \frac{3}{8} J_2 \left(n_s^2 + \mu_m \, n_m^2\right) \cos(2\,\theta), \tag{15.116}$$

where $U_{esm}^{(0)}$ is a constant. By analogy with Equation (15.101), as a consequence of the Earth's oblateness, the Sun and the Moon are able to exert a gravitational torque on the Earth; namely,

$$\boldsymbol{\tau} = -\frac{dU_{esm}}{d\theta}\left(-\sin\phi,\,\cos\phi,\,0\right)$$

$$= -\frac{3}{4} J_2 \left(n_s^2 + \mu_m \, n_m^2\right) M\,R^2 \sin(2\,\theta)\left(-\sin\phi,\,\cos\phi,\,0\right). \tag{15.117}$$

According to Equations (15.102) and (15.103), this torque causes the Earth's axis of rotation to precess about the normal to the ecliptic plane, in the opposite direction to the Sun's apparent orbit around the Earth, at the rate

$$\Omega_\phi = \frac{3}{2} \frac{\tilde{\epsilon}\,(n_s^2 + \mu_m\,n_m^2)}{\omega}\cos\theta. \tag{15.118}$$

This effect is known as *luni-solar precession*. The precession period in (sidereal) years is given by

$$T_\phi(\text{yr}) = \frac{n_s}{\Omega_\phi} = \frac{2\,T_s(\text{day})}{3\,\tilde{\epsilon}\,(1 + \mu_m/[T_m(\text{yr})]^2)\cos\theta}, \tag{15.119}$$

where $T_m(\text{yr}) = n_s/n_m = 0.00748$ is the Moon's (sidereal) orbital period in years. Given that $\tilde{\epsilon} = 3.27 \times 10^{-3}$, $\theta = 23.44°$, $T_s(\text{day}) = 366.26$, and $\mu_m = 0.0123$, we obtain

$$T_\phi \simeq 25\,400\,\text{years} \tag{15.120}$$

(Yoder 1995). This prediction is fairly close to the observed precession period of 25 800 years (Yoder 1995). The main reason that our estimate is slightly inaccurate is because we have neglected to take into account the small eccentricities of the Earth's orbit around the Sun and the Moon's orbit around the Earth, and have also neglected the small inclination of the Moon's orbit to the ecliptic plane.

The point in the sky toward which the Earth's axis of rotation is directed is known as the *north celestial pole*. Currently, this point lies within about a degree of the fairly bright star Polaris, which is consequently sometimes known as the *north star* or *pole star*. See Figure 15.7. It follows that Polaris appears to be almost stationary in the sky, always lying due north, and can thus be used for navigational purposes. Indeed, mariners have relied on the north star for many hundreds of years to determine direction at sea. Unfortunately, because of the precession of the Earth's axis of rotation, the north celestial pole is not a fixed point in the sky, but instead traces out a circle, of angular radius 23.44°, about the north ecliptic pole, with a period of 25 800 years. See Figure 15.7. Hence, a few thousand years from now, the north celestial pole will no longer coincide with Polaris, and there will be no convenient way of telling direction from the stars.

The projection of the ecliptic plane onto the sky is called the *ecliptic circle*, and coincides with the apparent path of the Sun against the backdrop of the stars. The projection of the Earth's equator onto the sky is known as the *celestial equator*. The ecliptic is inclined at 23.44° to the celestial equator. The two points in the sky at which the ecliptic crosses the celestial equator are called the equinoxes, because night and day are equally long when the Sun lies at these points. Thus, the Sun reaches the vernal equinox on about March 20, and this traditionally marks the beginning of spring. Likewise, the Sun reaches the autumn equinox on about September 22, and this traditionally marks the beginning of autumn. However, the precession of the Earth's axis of rotation causes the celestial equator (which is always normal to this axis) to precess in the sky; it thus also causes the equinoxes to precess along the ecliptic. This effect is known as the *precession of the equinoxes*. The precession is in the opposite direction to the Sun's apparent motion around the ecliptic, and is of magnitude 1.4° per century. Amazingly, this minuscule effect was discovered by the ancient Greeks (with the help of ancient Babylonian observations). In about 2000 BCE, when the science of astronomy originated in ancient Egypt and Babylonia, the vernal equinox lay in the constellation Aries. See Figure 15.8. Indeed, the vernal equinox is still sometimes called the *first point of Aries* in astronomical texts. About 90 BCE, the vernal equinox moved into the constellation Pisces, where it still remains. The equinox will move into the constellation Aquarius (marking the beginning of the much heralded "Age of Aquarius") in about 2600 CE. Incidentally, the position of the vernal equinox in the sky is of great

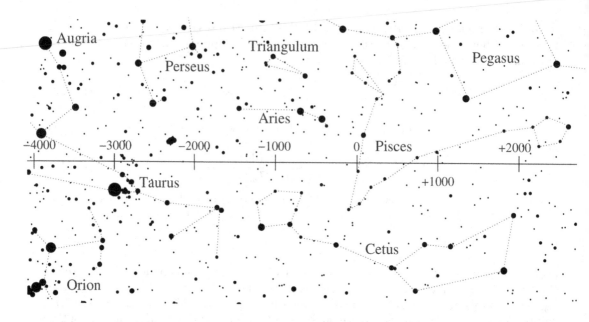

Figure 15.8 Path of the vernal equinox against the backdrop of the stars as a consequence of the precession of the equinoxes (calculated assuming constant precessional speed and obliquity). Numbers indicate years relative to start of common era. (Reproduced from Fitzpatrick 2012. Courtesy of Cambridge University Press.)

significance in astronomy, because it is used as the zero of celestial longitude (much as the Greenwich meridian is used as the zero of terrestrial longitude).

15.9 EXERCISES

15.1 Imagine a straight tunnel passing through the center of the Earth, which is regarded as a sphere of radius R and uniform mass density. A particle is dropped into the tunnel from the surface.

   (a) Show that the particle undergoes simple harmonic motion at the angular frequency $\omega = \sqrt{g/R}$, where g is the gravitational acceleration at Earth's surface.

   (b) Estimate how long it takes the particle to reach the other end of the tunnel. [Ans: 84.4 minutes.]

   (c) Estimate the speed of the particle as it passes through the center of the Earth. [Ans: 7.9 km/s.]

   (d) Obviously, a real tunnel cannot pass through the Earth's molten core. Assuming that the tunnel is smooth (i.e., ignoring friction), show that motion is simple harmonic even if the tunnel does not pass through the center of the Earth, and that the travel time is the same as before.

   [From Ingard 1988.]

15.2 Treating the Earth as a uniform-density, liquid spheroid, and taking rotational flattening into account, show that the variation of the surface acceleration, g, with terrestrial latitude, λ, is

$$g \simeq \frac{GM}{R^2} + \frac{1}{6}\Omega^2 R - \frac{5}{4}\Omega^2 R \cos^2 \lambda.$$

Here, M is the terrestrial mass, R the mean terrestrial radius, and Ω the terrestrial diurnal angular velocity. [From Fitzpatrick 2012.]

15.3 The Moon's mean orbital period about the Earth is approximately 27.32 days, and its orbital motion is in the same direction as the Earth's diurnal rotation (whose period is 24 hours). Use this data to show that, on average, high tides at a given point on the Earth occur every 12 hours and 27 minutes. [From Fitzpatrick 2012.]

15.4 Demonstrate that the mean time interval between successive spring tides is 14.76 days. [From Fitzpatrick 2012.]

15.5 Suppose that the Earth is modeled as a completely rigid sphere that is covered by a shallow ocean of negligible density.

   (a) Demonstrate that the tidal elongation of the ocean layer due to the Moon is

   $$\frac{\Delta R}{R} = \frac{3}{2}\frac{m'}{m}\left(\frac{R}{a}\right)^3,$$

   where m is the mass of the Earth, m' the mass of the Moon, R the radius of the Earth, and a the radius of the lunar orbit.

   (b) Show that $\Delta R = 0.53\,\mathrm{m}$.

   (c) Show that the tidal elongation of the ocean layer due to the Sun is such that $\Delta R = 0.25\,\mathrm{m}$.

   [From Fitzpatrick 2012.]

15.6 Estimate the tidal elongation of the Sun due to the Earth. [Ans: $3 \times 10^{-4}\,\mathrm{m}$.] [From Fitzpatrick 2012.]

15.7 The length of the *mean sidereal year*, which is defined as the average time required for the Sun to (appear to) complete a full orbit around the Earth, relative to the fixed stars, is 365.25636 days. The *mean tropical year* is defined as the average time interval between successive vernal equinoxes. Demonstrate that, as a consequence of the precession of the equinoxes, whose period is 25 772 years, the length of the mean tropical year is 20.4 minutes shorter than that of the mean sidereal year (i.e., 365.24219 days). [From Fitzpatrick 2012.]

Useful Mathematics

A.1   CALCULUS

$$\frac{d}{dx}\, x^n = n\, x^{n-1} \qquad (A.1)$$

$$\frac{d}{dx}\, \mathrm{e}^x = \mathrm{e}^x \qquad (A.2)$$

$$\frac{d}{dx}\, \ln x = \frac{1}{x} \qquad (A.3)$$

$$\frac{d}{dx}\, \sin x = \cos x \qquad (A.4)$$

$$\frac{d}{dx}\, \cos x = -\sin x \qquad (A.5)$$

$$\frac{d}{dx}\, \tan x = \frac{1}{\cos^2 x} \qquad (A.6)$$

$$\frac{d}{dx}\, \sin^{-1} x = \frac{1}{\sqrt{1-x^2}} \qquad (A.7)$$

$$\frac{d}{dx}\, \cos^{-1} x = -\frac{1}{\sqrt{1-x^2}} \qquad (A.8)$$

$$\frac{d}{dx}\, \tan^{-1} x = \frac{1}{1+x^2} \qquad (A.9)$$

$$\frac{d}{dx}\, \sinh x = \cosh x \qquad (A.10)$$

$$\frac{d}{dx}\, \cosh x = \sinh x \qquad (A.11)$$

$$\frac{d}{dx}\, \tanh x = \frac{1}{\cosh^2 x} \qquad (A.12)$$

$$\frac{d}{dx}\, \sinh^{-1} x = \frac{1}{\sqrt{x^2+1}} \qquad (A.13)$$

$$\frac{d}{dx}\, \cosh^{-1} x = \pm\frac{1}{\sqrt{x^2-1}} \qquad (A.14)$$

$$\frac{d}{dx}\, \tanh^{-1} x = \frac{1}{1-x^2} \qquad (A.15)$$

A.2   SERIES EXPANSIONS

Notation: $k! = k\,(k-1)\,(k-2)..2.1$, $f^{(n)}(x) = d^n f(x)/dx^n$.

$$f(x) = f(a) + \frac{(x-a)}{1!}\,f^{(1)}(a) + \frac{(x-a)^2}{2!}\,f^{(2)}(a) + \cdots \tag{A.16}$$

$$(1+x)^\alpha = 1 + \alpha\,x + \frac{\alpha\,(\alpha-1)}{2!}\,x^2 + \frac{\alpha\,(\alpha-1)\,(\alpha-2)}{3!}\,x^3 + \cdots \tag{A.17}$$

$$e^x = 1 + x + \frac{x^2}{2!} + \frac{x^3}{3!} + \cdots \tag{A.18}$$

$$\ln(1+x) = x - \frac{x^2}{2} + \frac{x^3}{3} - \frac{x^4}{4} + \cdots \tag{A.19}$$

$$\sin x = x - \frac{x^3}{3!} + \frac{x^5}{5!} - \frac{x^7}{7!} + \cdots \tag{A.20}$$

$$\cos x = 1 - \frac{x^2}{2!} + \frac{x^4}{4!} - \frac{x^6}{6!} + \cdots \tag{A.21}$$

$$\tan x = x + \frac{x^3}{3} + \frac{2\,x^5}{15} + \frac{17\,x^7}{315} + \cdots \tag{A.22}$$

$$\sinh x = x + \frac{x^3}{3!} + \frac{x^5}{5!} + \frac{x^7}{7!} + \cdots \tag{A.23}$$

$$\cosh x = 1 + \frac{x^2}{2!} + \frac{x^4}{4!} + \frac{x^6}{6!} + \cdots \tag{A.24}$$

$$\tanh x = x - \frac{x^3}{3} + \frac{2\,x^5}{15} - \frac{17\,x^7}{315} + \cdots \tag{A.25}$$

A.3 TRIGONOMETRIC IDENTITIES

$$\sin(-\alpha) = -\sin\alpha \tag{A.26}$$

$$\cos(-\alpha) = +\cos\alpha \tag{A.27}$$

$$\tan(-\alpha) = -\tan\alpha \tag{A.28}$$

$$\sin(\alpha \pm \pi/2) = \pm\cos\alpha \tag{A.29}$$

$$\cos(\alpha \pm \pi/2) = \mp\sin\alpha \tag{A.30}$$

$$\sin^2\alpha + \cos^2\alpha = 1 \tag{A.31}$$

$$\sin(\alpha \pm \beta) = \sin\alpha\,\cos\beta \pm \cos\alpha\,\sin\beta \tag{A.32}$$

$$\cos(\alpha \pm \beta) = \cos\alpha\,\cos\beta \mp \sin\alpha\,\sin\beta \tag{A.33}$$

$$\tan(\alpha \pm \beta) = \frac{\tan\alpha \pm \tan\beta}{1 \mp \tan\alpha\,\tan\beta} \tag{A.34}$$

$$\sin\alpha + \sin\beta = 2\,\sin\left(\frac{\alpha+\beta}{2}\right)\cos\left(\frac{\alpha-\beta}{2}\right) \tag{A.35}$$

$$\sin\alpha - \sin\beta = 2\,\cos\left(\frac{\alpha+\beta}{2}\right)\sin\left(\frac{\alpha-\beta}{2}\right) \tag{A.36}$$

$$\cos\alpha + \cos\beta = 2\,\cos\left(\frac{\alpha+\beta}{2}\right)\cos\left(\frac{\alpha-\beta}{2}\right) \tag{A.37}$$

$$\cos\alpha - \cos\beta = -2\,\sin\left(\frac{\alpha+\beta}{2}\right)\sin\left(\frac{\alpha-\beta}{2}\right) \tag{A.38}$$

$$\sin\alpha\,\sin\beta = \frac{1}{2}\left[\cos(\alpha-\beta) - \cos(\alpha+\beta)\right] \tag{A.39}$$

$$\cos\alpha\,\cos\beta = \frac{1}{2}\left[\cos(\alpha-\beta) + \cos(\alpha+\beta)\right] \tag{A.40}$$

$$\sin\alpha\,\cos\beta = \frac{1}{2}\left[\sin(\alpha-\beta) + \sin(\alpha+\beta)\right] \tag{A.41}$$

$$\sin(\alpha/2) = \pm\left(\frac{1-\cos\alpha}{2}\right)^{1/2} \tag{A.42}$$

$$\cos(\alpha/2) = \pm\left(\frac{1+\cos\alpha}{2}\right)^{1/2} \tag{A.43}$$

$$\tan(\alpha/2) = \pm\left(\frac{1-\cos\alpha}{1+\cos\alpha}\right)^{1/2} = \frac{1-\cos\alpha}{\sin\alpha} = \frac{\sin\alpha}{1+\cos\alpha} \tag{A.44}$$

$$\sin(2\alpha) = 2\,\sin\alpha\,\cos\alpha \tag{A.45}$$

$$\cos(2\alpha) = \cos^2\alpha - \sin^2\alpha = 2\cos^2\alpha - 1 = 1 - 2\sin^2\alpha \tag{A.46}$$

$$\sin^2\alpha = \frac{1}{2}\left[1 - \cos(2\alpha)\right] \tag{A.47}$$

$$\cos^2\alpha = \frac{1}{2}\left[1 + \cos(2\alpha)\right] \tag{A.48}$$

A.4 HYPERBOLIC IDENTITIES

$$\sinh(-\alpha) = -\sinh\alpha \tag{A.49}$$

$$\cosh(-\alpha) = +\cosh\alpha \tag{A.50}$$

$$\tanh(-\alpha) = -\tanh\alpha \tag{A.51}$$

$$\cosh^2\alpha - \sinh^2\alpha = 1 \tag{A.52}$$

$$\sinh(\alpha \pm \beta) = \sinh\alpha\,\cosh\beta \pm \cosh\alpha\,\sinh\beta \tag{A.53}$$

$$\cosh(\alpha \pm \beta) = \cosh\alpha\,\cosh\beta \pm \sinh\alpha\,\sinh\beta \tag{A.54}$$

$$\tanh(\alpha \pm \beta) = \frac{\tanh\alpha \pm \tanh\beta}{1 \pm \tanh\alpha\,\tanh\beta} \tag{A.55}$$

$$\sinh\alpha + \sinh\beta = 2\sinh\left(\frac{\alpha+\beta}{2}\right)\cosh\left(\frac{\alpha-\beta}{2}\right) \tag{A.56}$$

$$\sinh\alpha - \sinh\beta = 2\cosh\left(\frac{\alpha+\beta}{2}\right)\sinh\left(\frac{\alpha-\beta}{2}\right) \tag{A.57}$$

$$\cosh\alpha + \cosh\beta = 2\cosh\left(\frac{\alpha+\beta}{2}\right)\cosh\left(\frac{\alpha-\beta}{2}\right) \tag{A.58}$$

$$\cosh\alpha - \cosh\beta = 2\sinh\left(\frac{\alpha+\beta}{2}\right)\sinh\left(\frac{\alpha-\beta}{2}\right) \tag{A.59}$$

$$\sinh\alpha\,\sinh\beta = \frac{1}{2}\left[\cosh(\alpha+\beta) - \cosh(\alpha-\beta)\right] \tag{A.60}$$

$$\cosh\alpha\,\cosh\beta = \frac{1}{2}\left[\cosh(\alpha+\beta) + \cosh(\alpha-\beta)\right] \tag{A.61}$$

$$\sinh\alpha\,\cosh\beta = \frac{1}{2}\left[\sinh(\alpha+\beta) + \sinh(\alpha-\beta)\right] \tag{A.62}$$

$$\sinh(\alpha/2) = \pm\left(\frac{\cosh\alpha - 1}{2}\right)^{1/2} \tag{A.63}$$

$$\cosh(\alpha/2) = \left(\frac{\cosh\alpha + 1}{2}\right)^{1/2} \tag{A.64}$$

$$\tanh(\alpha/2) = \pm\left(\frac{\cosh\alpha - 1}{\cosh\alpha + 1}\right)^{1/2} = \frac{\cosh\alpha - 1}{\sinh\alpha} = \frac{\sinh\alpha}{\cosh\alpha + 1} \tag{A.65}$$

$$\sinh(2\alpha) = 2\sinh\alpha\,\cosh\alpha \tag{A.66}$$

$$\cosh(2\alpha) = \cosh^2\alpha + \sinh^2\alpha = 2\cosh^2\alpha - 1 = 2\sinh^2\alpha + 1 \tag{A.67}$$

A.5 COMPLEX IDENTITIES

$$e^{i\theta} = \cos\theta + i\sin\theta \tag{A.68}$$

$$\cos\theta = \frac{1}{2}\left(e^{i\theta} + e^{-i\theta}\right) \tag{A.69}$$

$$\sin\theta = \frac{1}{2i}\left(e^{i\theta} - e^{-i\theta}\right) \tag{A.70}$$

$$\cosh\theta = \frac{1}{2}\left(e^{\theta} + e^{-\theta}\right) \tag{A.71}$$

$$\sinh\theta = \frac{1}{2}\left(e^{\theta} - e^{-\theta}\right) \tag{A.72}$$

$$\cos(i\theta) = \cosh\theta \tag{A.73}$$

$$\sin(i\theta) = i\sinh\theta \tag{A.74}$$

$$\cosh(i\theta) = \cos\theta \tag{A.75}$$

$$\sinh(i\theta) = i\sin\theta \tag{A.76}$$

A.6 VECTOR IDENTITIES

$$\mathbf{a}\cdot\mathbf{b}\times\mathbf{c} = \mathbf{a}\times\mathbf{b}\cdot\mathbf{c} \tag{A.77}$$

$$\mathbf{a}\times(\mathbf{b}\times\mathbf{c}) = (\mathbf{a}\cdot\mathbf{c})\,\mathbf{b} - (\mathbf{a}\cdot\mathbf{b})\,\mathbf{c} \tag{A.78}$$

$$(\mathbf{a}\times\mathbf{b})\times\mathbf{c} = (\mathbf{c}\cdot\mathbf{a})\,\mathbf{b} - (\mathbf{c}\cdot\mathbf{b})\,\mathbf{a} \tag{A.79}$$

$$(\mathbf{a}\times\mathbf{b})\cdot(\mathbf{c}\times\mathbf{d}) = (\mathbf{a}\cdot\mathbf{c})\,(\mathbf{b}\cdot\mathbf{d}) - (\mathbf{a}\cdot\mathbf{d})\,(\mathbf{b}\cdot\mathbf{c}) \tag{A.80}$$

$$(\mathbf{a}\times\mathbf{b})\times(\mathbf{c}\times\mathbf{d}) = (\mathbf{a}\times\mathbf{b}\cdot\mathbf{d})\,\mathbf{c} - (\mathbf{a}\times\mathbf{b}\cdot\mathbf{c})\,\mathbf{d} \tag{A.81}$$

Bibliography

Den Hartog, J.P. 1961. *Mechanics*. Dover. New York NY.

Fitzpatrick, R. 2008. *Maxwell's Equations and the Principles of Electromagnetism*. Jones & Bartlett, Ascend Learning. Burlington MA.

Fitzpatrick, R. 2012. *An Introduction to Celestial Mechanics*. Cambridge University Press. Cambridge UK.

Fitzpatrick, R. 2019. *Oscillations and Waves: An Introduction*. 2nd Edition. Taylor & Francis. CRC Press. Boca Raton FL.

Fowles, G.R., and Cassiday, G.L. 2005. *Analytic Mechanics*. 7th Edition. Brooks/Cole, Thompson Learning. Belmont CA.

Goldstein, H., Poole, C., and Safko, J. 2001. *Classical Mechanics*. 3rd Edition. Addison-Wesley. San Francisco CA.

Ingard, K.U. 1988. *Fundamentals of Waves and Oscillations*. Cambridge University Press. Cambridge UK.

Lamb, H. 1942. *Dynamics*. 2nd Edition. Cambridge University. Press Cambridge UK.

Newton, I. 1687. *Philosophiae Naturalis Principia Mathematica*. London.

Pain, H.J. 1999. *The Physics of Vibrations and Waves*. 5th Edition. J. Wiley & Sons. Chichester UK.

Thornton, S.T., and Marion, J.B. 2004. *Classical Dynamics of Particles and Systems*. 5th Edition. Brooks/Cole, Thompson Learning, Belmont CA.

Wikipedia contributors 2021. *Wikipedia: The Free Encyclopedia*. http://en.wikipedia.org.

Yoder, C.F. 1995. *Astrometric and Geodetic Properties of Earth and the Solar System*, in *Global Earth Physics: A Handbook of Physical Constants*, Ahrens, T. (ed.). American Geophysical Union. Washington DC.

Index

Printed in the United States
by Baker & Taylor Publisher Services